EQUILIBRIUM STATISTICAL MECHANICS

EQUILIBRIUM STATISTICAL MECHANICS

Second Edition

FRANK C. ANDREWS

Merrill College, University of California, Santa Cruz

A WILEY-INTERSCIENCE PUBLICATION

John Wiley & Sons

NEW YORK · LONDON · SYDNEY · TORONTO

Library of Congress Cataloging in Publication Data

Andrews, Frank C

Equilibrium statistical mechanics.

"A Wiley-Interscience publication."

Companion vol. to the author's Thermodynamics: principles and applications.

Bibliography: p. 000

1. Statistical mechanics. I. Title.

QC174.8.A534 1974 530.1'32 74-17197

ISBN 0-471-03123-2

Printed in the United States of America

10 9 8 7 6 5 4 3 2

TO MY PARENTS

PREFACE TO THE SECOND EDITION

The first edition of this book arose from my conviction that sound understanding of the basics of statistical mechanics was vital for almost all students of science and technology. After all, the problem confronted by statistical mechanics is the rationalization of the macroscopic behavior of matter through the knowledge that the matter consists of discrete particles. Science students spend much time studying the molecules and much time studying macroscopic properties, but very little time linking the two studies. One reason is the imposing mathematical and scientific sophistication demanded by most statistical mechanics books. I felt that those requirements were excessive and that even at the level of the usual undergraduate physical chemistry course, students could come to understand the basic tools of statistical mechanics, use those tools in a variety of simple applications, and work within a framework of definitions and viewpoints capable of direct extension all the way to the frontiers of statistical mechanical research. I also thought it would be useful to a student at any level who was coping with one of the complicated treatises on the subject to have a short book that concentrated on clear explanations of the fundamentals.

I was encouraged by the reception given that volume, but my own teaching from it convinced me that no book should have to experience a first edition! A decade of use by a variety of students and teachers showed me that the entire effort needed redoing—new organization, new content, new proofs, new applications, new problems, and new explanations. This volume is the result.

I am convinced that it is important for a person to keep thermodynamic argumentation separate from statistical mechanical. To this end, I have written a companion volume to this one: *Thermodynamics—Principles and Applications* (Wiley-Interscience, New York, 1971), referred to in the text simply as "Andrews," which I, at least, strongly recommend.

Statistical mechanics is by no means a finished subject. Many of the derivations in this book are new and contain the seed of potentially exciting research efforts. As the last chapter makes clear, some of the most basic questions, such as why are there liquids and what are their properties, cannot now be answered satisfactorily by statistical mechanics. We must all make our own order out of the welter of ideas and facts we find. In doing this we have the heritage from all those who have gone before to guide us. For here we are reaching, however hesitantly, toward the goal of scientific theory—to show the logical necessity of the behavior of the physical world from as simple a set of basic assumptions as possible.

These are powerful insights and tools statistical mechanics offers. I hope this book and its subject prove enjoyable to the readers and that their subsequent use of the skills acquired will always be for the benefit of the human race.

I wish to thank numerous students and teachers who have given me their encouragement and suggestions. I especially thank A. C. Andrews and Arthur D. Payton.

Frank C. Andrews

Santa Cruz, California
May 1974

CONTENTS

EQUILIBRIUM STATISTICAL MECHANICS

PART 1

BASIC THEORY

1 WHAT IS STATISTICAL MECHANICS?

It is hard to realize that an atomic or molecular picture of the ultimate structure of matter became generally accepted by the scientific community only in the latter half of the nineteenth century. True, such a picture had been cropping up among philosophical writings for several thousand years. But scientists had held back from adopting it. Their desire was to unify and make order out of the endless, varied, apparently unrelated phenomena all about them by showing that everything would follow logically if certain simple hypotheses were true. Today's scientists have the same desire. Only after much development of the consequences of the molecular picture, many of which are presented in this book, did the atomic-molecular picture become a compelling one. It is certainly one of the most successful creative constructions of the scientific community, having tied together so many previously unrelated theories and phenomena and led to so many remarkable new realms of study and technology.

Of course, a microscopic particulate theory of matter is not thrust on people, because almost all the physical phenomena ever studied are *macroscopic*. That is, they involve matter in bulk, say, a chunk of metal, a beaker of liquid, or a tank of gas. The quantitative relationships coupling the macroscopic properties of matter were studied by such disciplines as mechanics, electricity and magnetism, and thermodynamics. Thermodynamics, for example, links the many observable properties of bulk matter, knowledge of a few thus permitting calculation of many others. The "truth" of thermodynamics rests not in the degree of refinement of some model but in its never failing to give valid answers to problems involving macroscopic systems.* Except perhaps in discussing the third law, thermodynamics takes no account of the atomic and molecular structure of matter. It is simply descriptive of what is observed and is not based on a model in which one set of phenomena are "explained" in terms of another that is more tangible or understandable. For such a causal

*For further discussion, see F. C. Andrews, *Thermodynamics—Principles and Applications* (Wiley-Interscience, New York, 1971) pp. 1–72.

"explanation" of the laws of thermodynamics, one must look to the properties and mechanics of the particles comprising the macroscopic system. Then one would expect to be able to show that a vast number of molecules rattling around and interacting among themselves according to some appropriate mechanical laws would give rise to the macroscopic phenomena we see. Before 1900 the mechanical laws governing the molecular motions were limited to classical mechanics, but the development of quantum mechanics during the period 1901 to 1926 brought much greater precision to the description. Of course, that has simply pushed the descriptive explanation back to the quantum mechanics, since there is no further model by which quantum mechanics becomes intuitively understandable.

Statistical mechanics can be viewed as the discipline bridging the gap between the microscopic world of molecules, atoms, electrons, and photons, and the macroscopic world of thermodynamics and the properties of materials. Given the quantitative properties of the molecules and how they interact, statistical mechanics sets out to calculate what macroscopic behavior to predict as a function of the material comprising the system and the constraints (e.g., volume and temperature) on the system. Sometimes the opposite tack is taken: given the macroscopic behavior, statistical mechanics is used to shed light on the properties the molecules must have. The magnitude of the many-particle problem confronting statistical mechanics can be appreciated by realizing just how many molecules are in macroscopic objects. If each molecule in a mole were the size of a small dried pea, there would be enough peas to cover the United States to a depth of several hundred feet. When we consider that it has proved impossible in either quantum or classical mechanics to solve the equations that give the trajectories of *just three* mutually interacting bodies, the problem here, involving 6×10^{23} bodies, is almost laughable. In fact, if we just wrote down the x, y, and z coordinates of each molecule, a thousand to the page and a thousand per minute, it would take a million times as long to write them as the estimated age of the earth, and a freight train loaded with the pages would stretch to the sun and back about 40,000 times!

It has been a long, hard process even to establish a language and a means of approaching the problem, and that process is still going on. The result, statistical mechanics, is a mixture of mechanics and probability theory. In understanding it, it is important to keep clear what part is mechanics and what part probability. Confusion between these has resulted in a century of misconceptions and quarrels. The early successes of statistical mechanics hastened the acceptance of molecular theory; later successes have shed light on countless phenomena. The many still insurmountable difficulties, however, make statistical mechanics far less inclusive and less accurate than one would wish. For example, the whole realm of dense gases and liquids and the liquid-vapor phase transition is understood only on a qualitative level at best. Precise statistical mechanical treatments of these important substances remain beyond the talents of the best theoreticians.

The physicists whose successes established the foundations of statistical mechanics in the scientific world were J. Clerk Maxwell of Cambridge and Ludwig Boltzmann of Vienna. These men did their major work between approximately 1860 and 1900. The entire field was carefully studied by J. Willard Gibbs of Yale, whose brilliant synthesis and creative exposition published in 1902 brought order and stability to the foundations laid by Maxwell and Boltzmann. Since Gibbs, statistical mechanics has attracted many of the most brilliant and creative physicists and physical chemists, yet the results are still a long way from the ultimate goal.

PROBLEMS

1-1. On what do you base your *personal* belief in the atomic-molecular structure of matter?

1-2. Estimate the number of atoms that are rubbed off the sole of your shoe with every step you take. Make whatever reasonable assumptions are necessary to obtain your result.

2 PROBABILITY THEORY

Natural and social scientists must express many of their most powerful quantitative relationships in terms of probabilities rather than determined quantities. For example, it is much more useful to know that the velocities of gas molecules are distributed according to Maxwell's velocity distribution than to know, say, $pV = nRT$, although $pV = nRT$ and many other useful results can be derived from the velocity distribution under the appropriate conditions.

In this section we consider the language of probability theory, in which not only statistical mechanics but also a great many other scientific results are expressed. Probability theory is a well-defined branch of mathematics. Like any other branch of mathematics, it is a self-consistent way of defining and thinking about certain idealizations. To the scientist, mathematics is simply one of his logical tools—broadly speaking, the logic of quantity. When the scientist uses mathematics, he simply employs a logical method of examining the consequences of a set of assumptions. Whenever, in science, the mathematical conclusions seem unbelievable, we know that the assumptions must be modified to no longer imply the objectionable conclusions.

Probability theory is misused or vaguely understood by too many scientists. One is handicapped for life in thinking and understanding if he or she does not feel at home with it. We try here to present the theory, divorced from specific content, in a way that will make it relatively easy to understand.

ENSEMBLES AND PROBABILITIES

Probability theory treats the properties of a completely abstract idealization we call an ensemble. *An* **ensemble** *is a collection of members, each of which has certain interesting properties.* Depending on the problem, the ensemble may contain just a few members, many members, or an infinite number of members. Also depending on the problem, the members of the ensemble may be characterized by any number of properties, as appropriate.

Consider as an ensemble a certain hypothetical collection of cats. The cats in the ensemble do not actually exist; we just imagine their existence together, and we imagine their properties. These might be color, sex, age at last birthday, number of teeth, weight, length. Here, color and sex are qualitative properties, rather than quantitative, characterizing each cat. Age and number of teeth are quantitative properties that take only discrete values. Weight and length are continuous parameters that characterize each cat. Probability theory easily copes with all three kinds of property; in science, however, one is most likely to study the quantitative properties rather than the qualitative.

The **probability** *P of a certain property is defined to be the fraction of members of the ensemble having that property:*

$$P(\text{property}) \equiv \frac{\text{number of members of ensemble with property}}{\text{total number of members of ensemble}}. \quad (2\text{-}1)$$

If the property of interest is labeled i, Eq. 2-1 can be rewritten

$$\boxed{P_i \equiv \frac{n_i}{n},} \quad (2\text{-}2)$$

where n_i is the number of members of the ensemble having property i and n is the total number of members of the ensemble. Clearly, Eq. 2-2 agrees with our intuitive ideas about the likelihood of selecting a member at random from the ensemble and finding it to possess property i. That is of course the basis for the definition of probability chosen. But probabilities are functions of ensembles and are defined mathematically by Eq. 2-2; any hookup between the mathematical theory and likelihoods in the "real world" must arise from a careful investigation that goes beyond mere "probability theory."

The probability P_{30} of a cat with 30 teeth is defined as the fraction of the cats in the ensemble that have 30 teeth; that is, as the number n_{30} of such cats in the ensemble divided by the total number n of cats it contains.

Most of the mathematical properties of probabilities follow directly from the definition, Eq. 2-2. The rest of this chapter is devoted to those properties.

The probability of a male cat is defined as the fraction of the cats in the ensemble that are male; that is, as the number of male cats in the ensemble divided by the total number of cats it contains.

If property i occurs in no members of the ensemble, its probability is zero:

$$P_i = \frac{n_i}{n} = \frac{0}{n} = 0. \tag{2-3}$$

If property i appears in all members of the ensemble, its probability is unity:

$$P_i = \frac{n_i}{n} = \frac{n}{n} = 1. \tag{2-4}$$

If all cats in the ensemble are alive, the probability of a dead cat is zero, $P_{\text{dead}} = 0$. The probability of a living cat is unity, $P_{\text{living}} = 1$.

Properties i and j are said to be **mutually exclusive** if no member of the ensemble can have both properties i and j. For mutually exclusive properties, the probability of *either* property i *or* property j is the fraction of members of the ensemble having either property:

$$P_{i \text{ or } j} = \frac{n_{i \text{ or } j}}{n} = \frac{n_i + n_j}{n} = \frac{n_i}{n} + \frac{n_j}{n} = P_i + P_j. \tag{2-5}$$

Thus the probability of one of a set of mutually exclusive properties is the sum of the probabilities of the individual properties.

Since no cat can have two different numbers of teeth, tooth numbers are mutually exclusive properties of cats. Therefore, the probability $P_{29 \text{ or } 30}$ of a cat's having either 29 or 30 teeth is the probability P_{29} of 29 plus the probability P_{30} of 30: $P_{29 \text{ or } 30} = P_{29} + P_{30}$.

If P_i is the probability of property i, then $1 - P_i$ is the probability of *not* having property i:

$$P_{\text{not } i} = \frac{n_{\text{not } i}}{n} = \frac{n - n_i}{n} = 1 - P_i. \tag{2-6}$$

This follows because all members of the ensemble either do or do not have the property.

DISCRETE VARIABLES

Let us consider only properties that are quantitative, whose values are discrete, and for which each member of the ensemble has *some* value. Examples are number of teeth and age at last birthday of our cats. The different values of any such property are mutually exclusive, since no member of the ensemble can have two values for the same property (i.e., no cat can be both 13 and 14 years old). The probability that such a property has the value i is P_i, defined of course by Eq. 2-2.

A simple extension of Eq. 2-5 is the following: The sum of P_i over all possible values i of the property yields unity:

$$P_0 + P_1 + P_2 + \cdots \equiv \sum_i P_i = \sum_i \frac{n_i}{n} = \frac{n}{n} = 1. \tag{2-7}$$

This makes sense, for example, if the subscripts refer to ages of cats. Since $\Sigma_i P_i$ is the probability a cat has *some* value for its age, and we know they *all* have some value, the sum must yield the probability of a certainty, which is unity. The sum includes all the cats in the ensemble. A probability that satisfies Eq. 2-7 is said to be **normalized**, and Eq. 2-7 is often called the **normalization condition**. All sets of probabilities must satisfy Eq. 2-7. Often, the physical situation determines only that P_i is *proportional* to some function of i:

$$P_i \propto f(i); \qquad P_i = af(i). \tag{2-8}$$

This alone, however, is enough to fix the value of the **normalization constant** a, simply by requiring the probabilities of Eq. 2-8 to satisfy the normalization condition, Eq. 2-7:

$$P_{\text{certainty}} = \sum_i P_i = \sum_i af(i) = a \sum_i f(i) = 1. \tag{2-9}$$

Thus we have

$$a = \frac{1}{\Sigma_i f(i)}; \tag{2-10}$$

and

$$P_i = \frac{f(i)}{\Sigma_i f(i)}. \tag{2-11}$$

In Eq. 2-9, Eq. 2-8 was substituted into Eq. 2-7; the result, Eq. 2-11, is the normalized expression for P_i, derived solely from the physical knowledge contained in Eq. 2-8.

The ensemble establishes a value of P_i for each possible value i of the parameter being considered. Viewed in this way, P_i is a *function* of i in the usual sense of the word: P_i is a **function** of i if for every allowed i, a unique value of P_i is implied.

Suppose that each member of the ensemble is described by several discrete, quantitative properties. Call the values of these properties $i,j,\ldots,l$. The ensemble defines not only the simple probabilities for each *single* property in the form of Eq. 2-2,

$$P_i \equiv \frac{n_i}{n}; \qquad P_j \equiv \frac{n_j}{n}; \;\cdots;\; P_l = \frac{n_l}{n}, \tag{2-12}$$

but it also defines all the so-called *joint* probabilities of two or more different properties. The **joint probability** P_{ij} for the value i for the first property and j for the second is the fraction of members of the ensemble that have not only the value i for the first but also the value j for the second:

$$P_{ij} \equiv \frac{n_{i \text{ and } j}}{n}. \tag{2-13}$$

Thus for a member of the ensemble to count toward the numerator in Eq. 2-13, it must have the right values for both properties under consideration, i for the first and j for the second. The joint two-property probability P_{ij} is a function of the values i and j of the two properties.

Table 2-1. Joint Probability for Age and Number of Teeth of Cats, Based on a Hypothetical Ensemble

	Age						Sum
Number of Teeth	1	2	3	4	5	6	Total of Row
28	$P_{1,28}$	$P_{2,28}$	$P_{3,28}$	$P_{4,28}$	$P_{5,28}$	$P_{6,28}$	P_{28}
29	$P_{1,29}$	$P_{2,29}$	$P_{3,29}$	$P_{4,29}$	$P_{5,29}$	$P_{6,29}$	P_{29}
30	$P_{1,30}$	$P_{2,30}$	$P_{3,30}$	$P_{4,30}$	$P_{5,30}$	$P_{6,30}$	P_{30}
Sum of Column	P_1	P_2	P_3	P_4	P_5	P_6	

Consider Table 2-1, which shows the joint probabilities for age and numbers of teeth for a hypothetical ensemble in which all cats are between 1 and 6 years of age and have between 28 and 30 teeth. Each entry is the fraction of the cats in the ensemble which have that particular combination of age and number of teeth. The functional dependence of P_{ij} on i and j is determined from the table, which of course is based on the ensemble.

Consider the formal relationship between P_{ij} and P_i and P_j. How might one obtain P_i or P_j as function of i or j, knowing the joint probability P_{ij} as function of i and j? Our cat example might help visualize this. The table gives P_{ij}; suppose we want the probability P_1 of a 1-year old cat (i.e., the fraction of cats that are 1 year old regardless of how many teeth they have). We could count all the 1-year-old cats in the ensemble by summing all those in the 1-year-old column of the table. That would include all 1-year-olds, regardless of their teeth:

$$P_1 = \frac{n_1}{n} = \frac{n_{1,28} + n_{1,29} + n_{1,30}}{n} = P_{1,28} + P_{1,29} + P_{1,30}. \tag{2-14}$$

Similarly, we could find the probability P_{30} of a 30-toothed cat by summing across the last row of the table, thus counting all 30-toothed cats in the ensemble, regardless of age:

$$\begin{aligned} P_{30} &= \frac{n_{30}}{n} = \frac{n_{1,30} + n_{2,30} + n_{3,30} + n_{4,30} + n_{5,30} + n_{6,30}}{n} \\ &= P_{1,30} + P_{2,30} + P_{3,30} + P_{4,30} + P_{5,30} + P_{6,30}. \end{aligned} \tag{2-15}$$

Thus, in general, a single-property probability is obtained from a joint probability by summing the latter over all values of the properties not of interest:

$$P_i = \sum_j P_{ij}; \qquad P_j = \sum_i P_{ij}. \tag{2-16}$$

The ideas of Eq. 2-16 are much used. Probabilities like P_i and P_j are said to be of *lower order* than joint probabilities like P_{ij} from which they can be obtained. More information about the ensemble is obtained by knowing P_{ij} as a function of i and j than by knowing P_i as a function of i and P_j as a function of j separately. The process of finding a lower order probability from one of higher order, as in Eq. 2-16, is often called *reducing* the higher order probability, and the lower order probabilities thus obtained are called **reduced probabilities**. This usage of the word "reduced" differs from that sometimes employed in referring to dimensionless variables as "reduced variables."

Suppose each member of the ensemble is described *completely* (at least, so far as we are concerned) by a set of s discrete, quantitative properties $i,j,k,\ldots,s$. In this case, the *ultimate* in joint probabilities is the so-called

complete probability, which gives the fraction of members that have each possible combination of values for all s properties:

$$P_{ijk\cdots s} \equiv \frac{n_{i \text{ and } j \text{ and } k \text{ and } \cdots \text{ and } s}}{n}. \tag{2-17}$$

It is a function of the values of all the s properties that characterize a member of the ensemble. If it is known as a function of these properties, one has *complete* knowledge about the ensemble. That is, he knows the relative magnitudes of the numerator of Eq. 2-17 for all possible combinations of values of the s properties. This enables him to construct a replica of the ensemble completely, since the relative number of members of each type is known.

Quite evidently, all other probabilities relating to the ensemble may be formed by reducing the complete probability, (i.e., by summing it over all values of the properties that are not of interest). In particular,

$$P_j = \sum_{i,k,\ldots,s} P_{ijk\cdots s}; \qquad P_{ij} = \sum_{k,\ldots,s} P_{ijk\cdots s}. \tag{2-18}$$

If the properties of interest have the values we are looking for, that member of the ensemble contributes to the probability we are concerned with, whatever values the other properties may have.

Any two properties i and j are said to be explicitly **correlated** whenever knowledge of one affects the probability distribution for values of the other, that is, whenever the dependence of P_i on the value of i is altered by knowing the value of j. The properties are **uncorrelated**, then, whenever P_i is the same function of i, whatever value property j might be known to have. This means that the same dependence of P_i on the value of i as in the entire ensemble would also be found using a smaller ensemble containing only those members that have a particular value of j. That is, if i and j are uncorrelated,

$$P_i \equiv \frac{n_i}{n} = \frac{n_{i \text{ and } j}}{n_j} \qquad \text{for all values of } i \text{ and } j. \tag{2-19}$$

In Prob. 2-10 it is proved that Eq. 2-19 implies the analogous equation

$$P_j \equiv \frac{n_j}{n} = \frac{n_{i \text{ and } j}}{n_i} \qquad \text{for all values of } i \text{ and } j. \tag{2-20}$$

If properties i and j are uncorrelated, P_i and P_j are said to be *independent* probabilities.

It follows from the definition that if i and j are uncorrelated properties, the joint probability P_{ij} is the product of P_i and P_j:

$$P_{ij} \equiv \frac{n_{i \text{ and } j}}{n} = \frac{n_i}{n} \frac{n_{i \text{ and } j}}{n_i} \tag{2-21}$$

$$= \frac{n_i}{n} \frac{n_j}{n} = P_i P_j; \tag{2-22}$$

$$\boxed{P_{ij} = P_i P_j; \qquad P_{ijk \cdots s} = P_i P_j P_k \cdots P_s,} \quad \text{(uncorrelated properties)} \tag{2-23}$$

where in Eq. 2-21 the numerator and denominator were both multiplied by n_i and in Eq. 2-22 use was made of Eq. 2-20.

So long as explicit knowledge of i does not change the probability distribution for values of j, P_{ij} factors into $P_i \times P_j$, even if i and j have a common cause that implies coupling between the properties. In scientific use of probability theory, one routinely looks for independent properties to factor the joint probability function. All the dependence of the joint probability on each property is contained in its own term in the product. If properties are not independent, their correlation prevents factoring the joint probability in such a way as to separate its dependence on the various properties.

If the number of teeth a cat had were completely independent of the cat's age, one would expect the probability for age and the probability for teeth to be independent of each other. The joint probability would factor. If this were the case, then in Table 2-1, for example, the fraction of cats with 30 teeth P_{30} in the entire ensemble would be the same as the fraction of 1-year-old cats with 30 teeth:

$$P_{30} = \frac{P_{1,30}}{P_{1,28} + P_{1,29} + P_{1,30}} = \frac{P_{1,30}}{P_1}. \tag{2-24}$$

That is, $P_{1,30}$ bears the same relationship to the sum of the 1-year-old column as the sum of the 30-toothed row bears to the sum of all the probabilities. Similarly, the fraction of 5-year-old cats P_5 in the total ensemble would be the same as the fraction of 28-toothed cats that are 5 years old:

$$P_5 = \frac{P_{5,28}}{P_{1,28} + P_{2,28} + P_{3,28} + P_{4,28} + P_{5,28} + P_{6,28}} = \frac{P_{5,28}}{P_{28}}. \tag{2-25}$$

That is, $P_{5,28}$ bears the same relationship to the sum of the 28-toothed row as the sum of the 5-year-old column bears to the sum of all the probabilities. Clearly, for real cats this would be only approximately true, since by the age of 1, cats have stopped growing new teeth, and during the years from 1 to 6 they do in fact occasionally lose them; thus there is a slight correlation between age and teeth for real cats; knowledge that a cat is old skews the probability distribution for teeth somewhat in the direction of fewer teeth.

ENSEMBLE AVERAGES

Consider a quantitative, discrete property i, represented by an ensemble. Suppose we are interested in some function $g(i)$ of the value of this property. Each member would of course have its value for the function; suppose we wanted to know the *average* value of g over the entire ensemble. The standard way to find the average of g is to sum g over all the members of the ensemble and divide the result by n:

$$\bar{g}=\frac{1}{n}\sum_{\substack{\text{all members}\\ \text{of ensemble}}} g_{i,\text{member}}. \tag{2-26}$$

Ensemble averages of quantities are noted in this book by overbars. It is convenient to find the sum in Eq. 2-26 in the following way. Place the members of the ensemble having the same value for i in separate groups; the number of members in each group with a particular value is n_i. The contribution to the sum in Eq. 2-26 from each group n_i is $n_i g_i$, since g_i has the same value for all members in a group. We can find the sum by adding the contributions $n_i g_i$ for each such group, one group for each different value of i:

$$\bar{g}=\frac{1}{n}\sum_i n_i g_i, \tag{2-27}$$

$$\boxed{\bar{g}=\sum_i P_i g_i,} \tag{2-28}$$

where we used Eq. 2-2 in obtaining Eq. 2-28.

A direct extension of the proof of Eq. 2-28 yields the analogous expression for $\bar{g}$ when the function g depends on several properties, $i,j,k,\ldots$:

$$\boxed{\bar{g}=\sum_{i,j,k,\cdots} P_{ijk\cdots}\, g_{ijk\cdots}.} \tag{2-29}$$

Consider an ensemble consisting of 10 cats with ages 1, 2, 2, 3, 3, 3, 4, 4, 5, 6; thus the probabilities generated are $P_1=.1$, $P_2=.2$, $P_3=.3$, $P_4=.2$, $P_5=.1$, and $P_6 = .1$. The average age could be found directly, as in Eq. 2-26, to be $33/10 = 3.3$. It could also be found from Eq. 2-28 to be $.1 + .4 + .9 + .8 + .5 + .6 = 3.3$. The average of the square of the age is, using Eq. 2-28, $.1 + .8 + 2.7 + 3.2 + 2.5 + 3.6 = 12.9$. Note: the average of the squares is larger than the square of the average ($3.3^2 = 10.89$). This invariably is the case with any probability distribution except the trivial one in which all members of the ensemble have the same value for the property of interest. The proof of that statement is given below, Eq. 2-40.

Of course, just specifying $\bar{g}$ represents a considerable loss of information from that contained in knowing the value of g for each member of the ensemble. One often wants to convey some feeling for how closely the members' g's are clustered about $\bar{g}$. The amount by which g_i differs from the mean for a particular choice of i is given the symbol δg_i:

$$\delta g_i \equiv g_i - \bar{g}. \tag{2-30}$$

The ensemble average of the deviation from the mean is of course zero:

$$\overline{\delta g} = \sum_i P_i(g_i - \bar{g}) \tag{2-31}$$

$$= \sum_i P_i g_i - \bar{g} \sum_i P_i \tag{2-32}$$

$$\overline{\delta g} = \bar{g} - \bar{g} = 0, \tag{2-33}$$

where in the last term of Eq. 2-32 we recognized that $\bar{g}$ is a constant, independent of i, which means that it comes outside the summation. This result follows from the definition of an average—the positive deviations exactly balance the negative deviations.

To obtain a useful measure of the average deviation of g from $\bar{g}$, the signs of positive and negative deviations must somehow be omitted to ensure that they will not cancel. One could average the absolute value of the deviation $|\delta g|$ to find the so-called **mean deviation**:

$$\overline{|\delta g|} = \sum_i P_i |g_i - \bar{g}| \tag{2-34}$$

$$= 2 \sum_{\substack{i \\ g_i < \bar{g}}} (P_i \bar{g} - P_i g_i) \tag{2-35}$$

$$\overline{|\delta g|} = 2\bar{g} \sum_{\substack{i \\ g_i < \bar{g}}} P_i - 2 \sum_{\substack{i \\ g_i < \bar{g}}} P_i g_i. \tag{2-36}$$

In Eq. 2-35 since it is recognized that the average positive and negative deviations are identical, twice the absolute value of the negative ones was used. In Eq. 2-36, $\bar{g}$ was recognized as independent of i.

It is, however, much more common to consider the average of the square of δg, rather than Eq. 2-36:

$$\overline{\delta g^2} = \sum_i P_i(g_i - \bar{g})^2 \tag{2-37}$$

$$= \sum_i P_i g_i^2 - 2\sum_i P_i g_i \bar{g} + \sum_i P_i \bar{g}^2 \tag{2-38}$$

$$= \overline{g^2} - 2\bar{g}^2 + \bar{g}^2 \tag{2-39}$$

$$\overline{\delta g^2} = \overline{g^2} - \bar{g}^2. \tag{2-40}$$

This quantity is called the **mean square deviation** or the **variance**. The left-hand side of Eq. 2-40 is of course positive, since it is the average of a positive quantity. Thus the right-hand side must always be positive, which proves the assertion made above that $\overline{g^2}$ is always greater than $\bar{g}^2$.

The square root of $\overline{\delta g^2}$ is called the **root mean square deviation**, or the **standard deviation** σ:

$$\sigma = \sqrt{\overline{\delta g^2}} = \sqrt{\overline{g^2} - \bar{g}^2}\,, \tag{2-41}$$

and a good measure of how the values of g for the members of the ensemble are spread out around $\bar{g}$ is to examine the ratio $\sigma/\bar{g}$, a dimensionless measure that is very small for tightly clustered g's and becomes larger as the values of g are less tightly clustered about the mean.

Sometimes other measures of spread in a distribution are chosen. The so-called ***n*th moment about the mean** μ_n is defined as

$$\mu_n \equiv \sum_i P_i \delta g_i^{\,n}. \tag{2-42}$$

We have already seen that μ_0 is 1, μ_1 is 0, and μ_2 is the variance. Higher moments are sometimes calculated for specific reasons that need not concern us here.

We are often in the position of having a probability distribution P_i as function of i and wanting to convey some information about it, but not its form in detail. We commonly give $\bar{i}$, the mean value of i:

$$\bar{i}=\sum_i P_i i, \tag{2-43}$$

but there are other possible measures of the "central thrust" of the distribution. For example, the **median** value of i is the value equaled or exceeded by exactly half the members of the ensemble. If the central pair of members have different values of i, the mean of this pair is usually chosen as the median. Another example is the so-called **mode** of the distribution, or **most probable value** of i, defined to be the value of i for which P_i takes its largest value. Once we have specified the mean, median, or mode, we can go further and express the dispersion of the i values about this central one by giving the mean deviation, variance, standard deviation, or similar measure.

CONTINUOUS VARIABLES

When properties described by the ensemble take continuous values, the problem is slightly different. If the exact position of a particle in a one-dimensional box is the property of interest, for example, then in an ensemble of finite size, the probability of any particular position would be zero. This is because there are an uncountable infinity of positions in any finite length of the box, and the chance that a member of a finite ensemble has its particle in *any* predetermined exact position is utterly negligible. It is the wrong question to ask for the probability that a continuous variable has a particular value. It makes more sense to ask for the probability that the value of the property *lies between* two values, say, between x and $x+dx$. Here dx is a finite interval but one that can be made as small as desired. Of course to accommodate infinitesimal dx's, we may need to construct ensembles with infinite numbers of members, but that need cause no alarm. The **probability** of a value of the property between x and $x+dx$ is defined as the fraction of members of the ensemble whose values lie between x and $x+dx$.

Now, suppose we make dx small enough that the probability that the value lies between x and $x+dx$ is not much different from the probability that it lies between $x+dx$ and $x+2dx$, and so on. Thus if we doubled the size of dx, the probability would double; if we halved the size of dx, the probability would go down by half. Unless the ensemble is highly pathological, we should easily be able to reach sufficiently small dx's that the probability of a value between x and $x+dx$ becomes proportional to the size of dx (as long as this proportionality is not pushed so far that dx gets too big). The proportionality constant is called the **probability density** or **distribution function** $f(x)$:

$$\text{Probability value lies between } x \text{ and } x+dx \equiv f(x)\,dx. \tag{2-44}$$

Consider as an example the hypothetical probability distribution for ages of cats (to last birthday) previously introduced, and pictured in Fig. 2-1*a*. This could be the representation of an ensemble consisting of just 10 cats, and it is given in terms of the probability function P_i where i is age. Exactly the same information is contained in Fig. 2-1*b*, except the ordinate is changed to a probability density D. To find the probability that a cat's age lies between one year and the next, one must find the area under the curve between the two ages. That explains why the units of D are reciprocal years (year^{-1}), since the product of D and 1 year must be a probability, and probabilities are always dimensionless. Figure 2-1*c* actually contains more information than either Fig. 2-1*a* or 2-1*b*, since it analyzes the ensemble in terms of half-integral ages, rather than integral. There would have to be a lot more than 10 members (perhaps 100 or more) in any ensemble represented by Fig. 2-1*c*, since 12 different ages are represented by 12 different relatively complicated probabilities (fractions). For Figs. 2-1*c* and 2-1*b* to be compatible, the area under the two half-integral ages between n and $n+1$ in Fig. 2-1*c* must equal the area between n and $n+1$ in Fig. 2-1*b*. In Fig. 2-1*d* this procedure of analyzing the ensemble in terms of smaller age ranges is pushed to the limit of a continuous age variable. Here the rectangles $f\,dn$ are infinitesimally thick. For Figs. 2-1*d* and 2-1*c* to be compatible, the area under the curve in Fig. 2-1*d* between any two adjacent half-integral ages must equal the area under the comparable part of the curve in Fig. 2-1*c*.

The terminology *probability density* for f is a precise analogy with the familiar *mass density* ρ, defined by the statement that $\rho(x,y,z)\,dx\,dy\,dz$ is the mass contained in the volume element $dx\ dy\ dz$. Here the volume is chosen so small that even if ρ changes with position, its change is negligible over distances like dx, dy, or dz; thus the mass contained in the tiny volume is proportional to the size of the volume, and the definition makes sense.

All the relationships developed previously in this chapter for probabilities with discrete variables carry over directly to probability densities for continuous variables with the summations over the former replaced by integrations over the latter. This follows because P_i is a fraction and so is $f(x)\,dx$; summations over P_i are $\Sigma_i P_i$; summations over $f(x)\,dx$ are $\int f(x)\,dx$. Let us enumerate the analogous results.

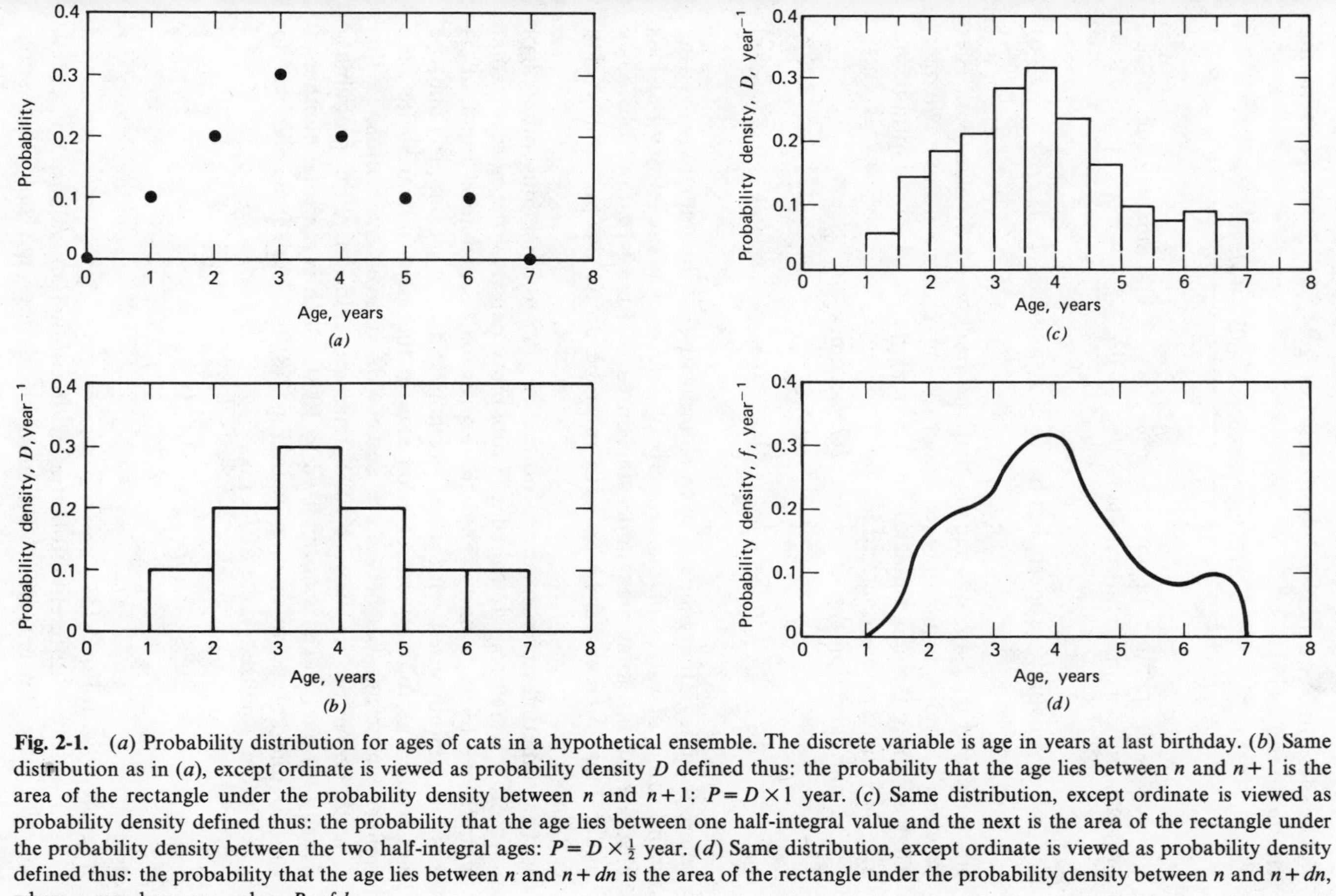

Fig. 2-1. (*a*) Probability distribution for ages of cats in a hypothetical ensemble. The discrete variable is age in years at last birthday. (*b*) Same distribution as in (*a*), except ordinate is viewed as probability density D defined thus: the probability that the age lies between n and $n+1$ is the area of the rectangle under the probability density between n and $n+1$: $P = D \times 1$ year. (*c*) Same distribution, except ordinate is viewed as probability density defined thus: the probability that the age lies between one half-integral value and the next is the area of the rectangle under the probability density between the two half-integral ages: $P = D \times \frac{1}{2}$ year. (*d*) Same distribution, except ordinate is viewed as probability density defined thus: the probability that the age lies between n and $n+dn$ is the area of the rectangle under the probability density between n and $n+dn$, where n may have any value: $P = f\,dn$.

The defining equation 2-2 becomes Eq. 2-44, which in terms of the ensemble is

$$f(x)\,dx = \frac{n(\text{value between } x \text{ and } x+dx)}{n}. \tag{2-45}$$

Since each member of the ensemble has only a single value of the property under consideration, the analog of Eq. 2-5 is

$$\text{Probability property lies between } x_1 \text{ and } x_2 = \int_{x_1}^{x_2} f(x)\,dx. \tag{2-46}$$

In terms of Fig. 2-1*d*, this means that the fraction of the cats with ages between n_1 and n_2 is just the area under the curve between those two ages. That is simply the sum (integral) of the areas $f(x)\,dx$ of all the infinitesimally thin slices between n_1 and n_2.

The normalization condition, Eq. 2-7, becomes

$$\int_{-\infty}^{\infty} f(x)\,dx = 1, \tag{2-47}$$

where the integration limits $-\infty$ to ∞ indicate that the integral is over the entire range of values of the property, whatever that range may be. This means that the total area under the curve in Fig. 2-1*d* (and obviously under Figs. 2-1*c* and 2-1*b* as well) must be unity, the probability of a certainty.

It is sometimes convenient to normalize $f(x)$ to something other than unity. If it were properly called a "probability density," the normalization would have to be to unity. Thus the expression "distribution function" is more commonly used with other normalizations. An example could be constructed as follows: out of 1000 cats in the local animal shelter, the *number* whose age lies between n and $n+dn$ is given by $F(n)\,dn$. In this case the value of $F(n)$ is just 1000 times the value of f, the probability density for age, and F is normalized to 1000, which is the total number of cats. If one knows, in analogy with Eq. 2-8, that $f(x)$ is *proportional* to $F(x)$, the normalized value of $f(x)$ is

$$f(x) = \frac{F(x)}{\int_{-\infty}^{\infty} F(x)\,dx}, \tag{2-48}$$

in analogy with Eq. 2-11.

If there are s different properties with continuous values, $x, y, z, \ldots, s$, described by the ensemble, the ensemble defines not only the simple probability densities $f(x), f(y), f(z), \ldots$, for each property, it also defines joint and multiple probability densities in analogy with Eq. 2-13. The complete distribution function $f_s(x,y,z,\ldots,s)$ for the ensemble gives full information about the ensemble. It is defined as follows: $f_s(x,y,z,\ldots,s)$ $dx\,dy\,dz \cdots ds$ is the fraction of members of the ensemble with the value of the first property between x and $x+dx$, and the value of the second property between y and $y+dy$, and so on. This is analogous to Eq. 2-17. The complete distribution function can be reduced to lower order distribution functions by integrating over all values of the properties in which one is not interested, in analogy to Eqs. 2-18:*

$$f(x) = \int_{-\infty}^{\infty} dy\,dz \cdots ds\, f_s(x,y,z,\ldots,s), \tag{2-49}$$

$$f_2(x,y) = \int_{-\infty}^{\infty} dz \cdots ds\, f_s(x,y,z,\ldots,s), \tag{2-50}$$

$$f(x) = \int_{-\infty}^{\infty} dy\, f_2(x,y). \tag{2-51}$$

If the properties given by x and y are uncorrelated, then the joint probability factors, in analogy with Eq. 2-23:

$$f_2(x,y) = f(x)f(y); \qquad \text{(uncorrelated properties)} \tag{2-52}$$

$$f_s(x,y,z,\ldots,s) = f(x)f(y)f(z)\cdots f(s). \qquad \text{(uncorrelated properties)} \tag{2-53}$$

The ensemble average of a function g of the value x of the property is given by

$$\bar{g} = \int_{-\infty}^{\infty} dx\, f(x)g(x), \tag{2-54}$$

in analogy with Eq. 2-28. If g is a function of several properties, its average

*In this book it is common to find each differential immediately behind its integral sign instead of after the integrand. This is useful in multiple integrals in which the different variables have different limits. It emphasizes the variable over which the integration occurs. Also, in multiple integrals, one integral sign is often made to serve for all the differentials that follow it if there is no question of the limits of integration.

is given by

$$\bar{g} = \int_{-\infty}^{\infty} dx\,dy\,dz \cdots ds\,f_s(x,y,z,\ldots,s)\,g(x,y,z,\ldots,s), \qquad (2\text{-}55)$$

in analogy with Eq. 2-29.

Expressions for mean deviation, mean square deviation (in particular, Eq. 2-40), and root mean square deviation (Eq. 2-41), and various moments about the mean, are completely analogous for continuous variables to those found for discrete. The *median* m of a distribution is easily defined with a continuous property to be the solution of either equation

$$\tfrac{1}{2} = \int_{-\infty}^{m} f(x)\,dx; \quad \tfrac{1}{2} = \int_{m}^{\infty} f(x)\,dx. \qquad (2\text{-}56)$$

CHANGE OF VARIABLE

Suppose we know how $f(x)$ depends on x; thus we know the probability that the value of the property lies between x and $x+dx$ as a function of x. Perhaps, however we are really interested in the probability $g(y)\,dy$ that the value of the function of x, $y(x)$, lies between y and $y+dy$. The problem is to relate $g(y)$ to $f(x)$. This is done immediately when we note that

$$f(x)\,dx = g(y)\,dy, \qquad (2\text{-}57)$$

so long as the interval dy is not independent but is that which accompanies the change dx in x. In other words, Eq. 2-57 is valid so long as

$$dy = \frac{dy}{dx}\,dx, \qquad (2\text{-}58)$$

which means Eq. 2-57 becomes

$$f(x)\,dx = g(y)\frac{dy}{dx}\,dx, \qquad (2\text{-}59)$$

which must be valid for all infinitesimal dx's. Thus the desired relationship between $f(x)$ and $g(y)$ is

$$f(x) = g[y(x)]\frac{dy}{dx} \quad \text{or} \quad g(y) = f[x(y)]\frac{dx}{dy}. \qquad (2\text{-}60)$$

The notation $g[y(x)]$ means the function $g(y)$ with y replaced by its value $y(x)$ in terms of x.

Example 2-1. Suppose the probability amplitude for values of the x-component of velocity of a molecule v_x is given by

$$\Phi(v_x) = \left(\frac{m}{2\pi kT}\right)^{1/2} e^{-mv_x^2/2kT}. \tag{2-61}$$

What is the definition of $\Phi(v_x)$? What is the most probable value of v_x? What is the value of $g(p_x)$, the probability amplitude for x-component of momentum ($p_x = mv_x$)? What is the definition of $g(p_x)$? What is the value of $h(K_x)$, the probability amplitude for kinetic energy due to motion in the x-direction ($K_x = \frac{1}{2}mv_x^2$)? What is the definition of $h(K_x)$?

Solution. The quantity $\Phi(v_x)$ is defined by the statement that $\Phi(v_x)\,dv_x$ is the probability that the x-component of velocity of a particle lies between v_x and $v_x + dv_x$. To find the most probable value of v_x, we find the value of v_x that gives $\Phi(v_x)$ its maximum value. This is done by setting the derivative of Φ equal to zero:

$$\frac{d\Phi}{dv_x} = \left(\frac{m}{2\pi kT}\right)^{1/2} e^{-mv_x^2/2kT}\left(-\frac{2mv_x}{2kT}\right) = 0. \tag{2-62}$$

Thus we have

$$v_{x,\,\text{most probable}} = 0. \tag{2-63}$$

Since the function $\Phi(v_x)$ has its maximum at $v_x = 0$, the most probable v_x is zero.

The value of $g(p_x)$ is found from Eq. 2-60:

$$g(p_x) = \Phi[v_x(p_x)]\frac{dv_x}{dp_x}, \tag{2-64}$$

$$g(p_x) = \left(\frac{m}{2\pi kT}\right)^{1/2} e^{-p_x^2/2mkT}\left(\frac{1}{m}\right). \tag{2-65}$$

The quantity $g(p_x)$ is defined by the statement that $g(p_x)\,dp_x$ is the probability that the x-component of momentum lies between p_x and $p_x + dp_x$.

The value of $h(K_x)$ is also found from Eq. 2-60:

$$h(K_x) = \Phi[v_x(K_x)]\frac{dv_x}{dK_x}, \tag{2-66}$$

$$h(K_x) = \left(\frac{m}{2\pi kT}\right)^{1/2} e^{-K_x/kT}\left(\frac{1}{2mK_x}\right)^{1/2}, \tag{2-67}$$

where the positive square root was chosen because probability amplitudes must by definition be positive.

The task of changing variables is slightly more complicated if more than one variable combines to form a new set of variables. If the old variables are $x,y,z,\ldots$, and the new are $r,s,t,\ldots$, the analog to Eq. 2-57 is

$$f(x,y,z,\cdots)dx\,dy\,dz\cdots = g(r,s,t,\cdots)dr\,ds\,dt\cdots, \tag{2-68}$$

so long as the multidimensional volume element $dr\,ds\,dt\cdots$ is not independent but is exactly what is caused by the changes $dx\,dy\,dz\cdots$. Here we must find the relationships between these volume elements, a relationship given formally by

$$dr\,ds\,dt\cdots = \frac{\partial(r,s,t,\cdots)}{\partial(x,y,z,\cdots)}dx\,dy\,dz\cdots, \tag{2-69}$$

where the so-called **Jacobian of the transformation*** $\partial(r,s,t,\cdots)/\partial(x,y,z,\cdots)$ is given by the determinant

$$\frac{\partial(r,s,t,\cdots)}{\partial(x,y,z,\cdots)} = \begin{vmatrix} \frac{\partial r}{\partial x} & \frac{\partial s}{\partial x} & \frac{\partial t}{\partial x} & \cdot & \cdot & \cdot \\ \frac{\partial r}{\partial y} & \frac{\partial s}{\partial y} & \frac{\partial t}{\partial y} & \cdot & \cdot & \cdot \\ \frac{\partial r}{\partial z} & \frac{\partial s}{\partial z} & \frac{\partial t}{\partial z} & \cdot & \cdot & \cdot \\ \cdot & \cdot & \cdot & \cdot & \cdot & \cdot \\ \cdot & \cdot & \cdot & \cdot & \cdot & \cdot \end{vmatrix}. \tag{2-70}$$

In terms of Jacobians, the multidimensional analog of Eq. 2-60 is

$$f(x,y,z) = g[r(x,y,z,\cdots), s(x,y,z,\cdots), t(x,y,z,\cdots),\cdots] \times \frac{\partial(r,s,t,\cdots)}{\partial(x,y,z,\cdots)}, \tag{2-71}$$

and a similar expression for obtaining g from f. Although these equations look forbidding, their application usually presents very little problem.

Example 2-2. Express the volume element in Cartesian coordinates $dx\,dy\,dz$ in terms of spherical coordinates r,θ, and ϕ, as in Fig. 2-2. Here we have

$$x = r\sin\theta\cos\phi, \qquad y = r\sin\theta\sin\phi, \qquad z = r\cos\theta. \tag{2-72}$$

*See any book on advanced calculus, for example, R. Courant, *Differential and Integral Calculus* (Blackie & Son Ltd., London, vol. II, 1936), pp. 156 and 254.

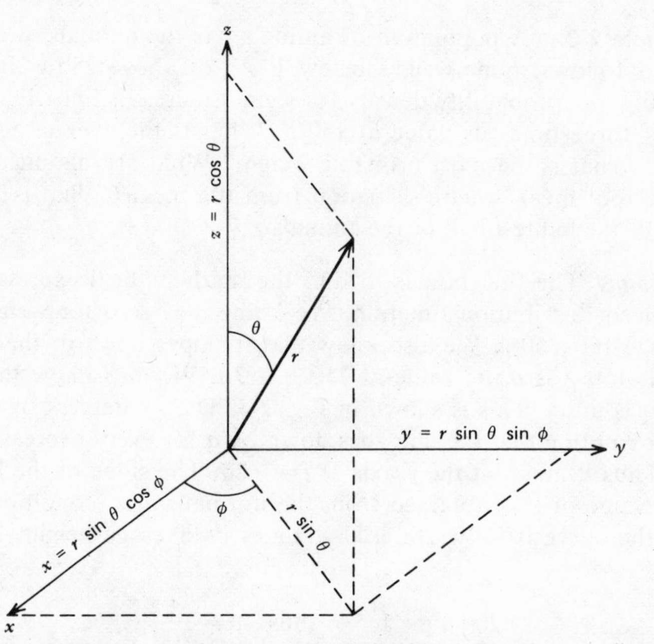

Fig. 2-2. Spherical coordinates r, θ, ϕ shown in relationship to Cartesian coordinates x,y,z.

Solution. One approach is to apply Eq. 2-69 directly, using Eqs. 2-72:

$$dx\,dy\,dz = \begin{vmatrix} \dfrac{\partial x}{\partial r} & \dfrac{\partial y}{\partial r} & \dfrac{\partial z}{\partial r} \\ \dfrac{\partial x}{\partial \theta} & \dfrac{\partial y}{\partial \theta} & \dfrac{\partial z}{\partial \theta} \\ \dfrac{\partial x}{\partial \phi} & \dfrac{\partial y}{\partial \phi} & \dfrac{\partial z}{\partial \phi} \end{vmatrix} dr\,d\theta\,d\phi, \qquad (2\text{-}73)$$

$$dx\,dy\,dz = \begin{vmatrix} \sin\theta\cos\phi & \sin\theta\sin\phi & \cos\theta \\ r\cos\theta\cos\phi & r\cos\theta\sin\phi & -r\sin\theta \\ -r\sin\theta\sin\phi & r\sin\theta\cos\phi & 0 \end{vmatrix} dr\,d\theta\,d\phi, \qquad (2\text{-}74)$$

$$dx\,dy\,dz = r^2\sin\theta\,dr\,d\theta\,d\phi, \qquad (2\text{-}75)$$

where the determinant was calculated with the recognition that $\sin^2\chi + \cos^2\chi = 1$.

Example 2-3. A population of animals has the animals' weights distributed as follows: none weighs below 150 g or above 250 g. Between 150 and 250 g, the probability density for weight increases linearly; thus at 250 g it has three times its value at 150 g. What is the average weight of an animal? What is the most probable weight? What are the mean deviation and the root mean square deviation from the mean? What is the average weight of the lightest half of the animals?

Solution. The first task is to find the mathematical expression for the normalized distribution function. We know it is zero for weights outside the range 150–250 g. We also know that it slopes upward; therefore, if its value at 150 g is a, its value at 250 g is $3a$. We also know that the area under it is unity. This is shown in Fig. 2-3. The y-intercept of the curve is found by noting that the line goes down by a for every decrease of 50 g in mass. Thus it must hit the y-axis at $f = -2a$. The slope of the line is $a/50$ g. The value of a is obtained from the normalization condition. The area under the curve is its width, 100 g, times its average height, $2a$, thus we have

$$200\, a \text{ g} = 1; \qquad \text{thus} \quad a = \frac{1}{200}\text{g}^{-1}. \tag{2-76}$$

Now, knowing the slope and y-intercept, we can write the normalized

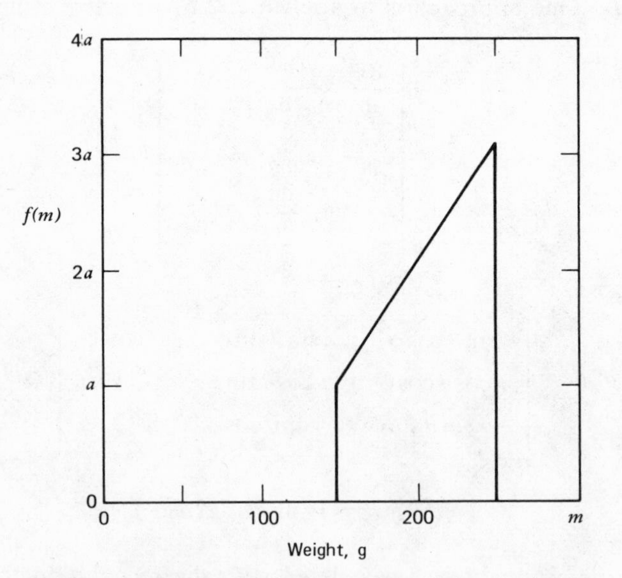

Fig. 2-3. Probability density for weights of lab animals.

probability density as follows:

$$f(m)=0, \qquad (m<150\text{ g})$$

$$f(m)=10^{-4}\text{ g}^{-2}m-10^{-2}\text{ g}^{-1}, \qquad (150\text{ g}\leqslant m\leqslant 250\text{ g})$$

$$f(m)=0 \qquad (m>250\text{ g}) \tag{2-77}$$

We now can compute $\overline{m}$ using Eq. 2-54:

$$\overline{m}=\int_0^\infty f(m)m\,dm, \tag{2-78}$$

$$=10^{-4}\text{ g}^{-2}\int_{150\text{ g}}^{250\text{ g}} m^2\,dm-10^{-2}\text{ g}^{-1}\int_{150\text{ g}}^{250\text{ g}} m\,dm, \tag{2-79}$$

$$=10^{-4}\text{ g}^{-2}\frac{m^3}{3}\bigg]_{150\text{ g}}^{250\text{ g}}-10^{-2}\text{ g}^{-1}\frac{m^2}{2}\bigg]_{150\text{ g}}^{250\text{ g}}, \tag{2-80}$$

$$\overline{m}=208.3333\text{ g}. \tag{2-81}$$

The most probable weight is the m that gives $f(m)$ its greatest value, which clearly is 250 g. The mean deviation is calculated from Eq. 2-36:

$$\overline{|\delta m|}=2\int_{\overline{m}}^{250\text{ g}} f(m)(m-\overline{m})\,dm, \tag{2-82}$$

$$=2\times10^{-4}\text{ g}^{-2}\int_{\overline{m}}^{250\text{ g}} m^2\,dm-2\times10^{-4}\text{ g}^{-2}\overline{m}\int_{\overline{m}}^{250\text{ g}} m\,dm,$$

$$-2\times10^{-2}\text{ g}^{-1}\int_{\overline{m}}^{250\text{ g}} m\,dm+2\times10^{-2}\text{ g}^{-1}\overline{m}\int_{\overline{m}}^{250\text{ g}} dm, \tag{2-83}$$

$$\overline{|\delta m|}=23.632\text{ g}. \tag{2-84}$$

The mean square deviation is obtained using Eq. 2-40:

$$\overline{\delta m^2}=\int_{150\text{ g}}^{250\text{ g}} f(m)m^2\,dm-\overline{m}^2, \tag{2-85}$$

$$=10^{-4}\text{ g}^{-2}\int_{150\text{ g}}^{250\text{ g}} m^3\,dm-10^{-2}\text{ g}^{-1}\int_{150\text{ g}}^{250\text{ g}} m^2\,dm-\overline{m}^2, \tag{2-86}$$

$$\overline{\delta m^2}=763.9061\text{ g}^2. \tag{2-87}$$

We note, as expected, that the mean square deviation is larger than the square of the mean deviation, since squaring quantities always gives additional weight to the largest of the quantities. To find the average weight of the lightest half of the animals, we note that these are *not* the same as all those with weights lower than the mean. It is the median, not the mean, that separates the ensemble into two parts of equal weight. The mean splits the first moment $\int f(m)m\,dm$ into two parts of equal absolute value, whereas the median splits the zeroth moment $\int f(m)\,dm$ into two such parts. We must first calculate the median M, using Eq. 2-56:

$$\tfrac{1}{2}=\int_{M}^{250\text{ g}} f(m)\,dm, \tag{2-88}$$

$$\tfrac{1}{2}=10^{-4}\text{ g}^{-2}\int_{M}^{250\text{ g}} m\,dm-10^{-2}\text{ g}^{-1}\int_{M}^{250\text{ g}} dm, \tag{2-89}$$

$$0.5\times10^{-4}M^2-10^{-2}M-0.125=0, \tag{2-90}$$

$$M=211.8034. \tag{2-91}$$

This shows that the median for this ensemble is slightly greater than the mean. Now, to find the average weight of the lightest half of the animals, we must normalize the probability density for the range 150 g to M. This is easy, since we know from Eq. 2-88 that the expression given by Eq. 2-77 is normalized in this range to $\frac{1}{2}$ and the probability density normalized to unity is therefore just twice the value in Eq. 2-77. The average weight of the lightest half is now immediately obtained:

$$\overline{m}_{\text{light}}=2\times10^{-4}\text{ g}^{-2}\int_{150\text{ g}}^{211.8034\text{ g}} m^2\,dm-2\times10^{-2}\text{ g}^{-1}\int_{150\text{ g}}^{211.8034\text{ g}} m\,dm, \tag{2-92}$$

$$\overline{m}_{\text{light}}=184.836\text{ g}. \tag{2-93}$$

All these calculations are made tolerable only if one has access to a calculating machine or electronic slide rule.

PROBLEMS

2-1. Since we do not know where a particle is in a one-dimensional box of length L, we assign it uniform probability density from $x=0$ to $x=L$, to reflect our ignorance. What is the normalized probability density for position, the average position, and the ratio of the root mean square deviation in position to the average position?

2-2. The probability density $f(t)$ for age t of a population of laboratory rats falls off to zero uniformly from 0 to 25 weeks. Express $f(t)$ mathematically; find the average age of a rat; find the median age of a rat. If the colony consists of 10,000 rats, how many newborns must be obtained daily?

2-3. An ant lives on the bottom of a 3×5-in. box. It is three times as likely to be somewhere in its feeding region, a circle of 1-inch diameter, as it is to be in all the rest of the total area of the box. (*a*) What is the probability it will be in the feeding circle? (*b*) What is the probability it will be in a circle of 1-inch diameter which does not overlap the feeding circle?

2-4. The probability density for speeds (magnitude of velocity) v of gas molecules is proportional to $v^2\exp[-mv^2/2kT]$. Normalize the distribution function and find the average speed. Use the integral tables in Appendix B.

2-5. The joint distribution function for x, y, and z components of velocity, v_x, v_y, and v_z for gas molecules at equilibrium is proportional to $\exp[-m(v_x^2+v_y^2+v_z^2)/2kT]$. Normalize the joint distribution function. Is there correlation between the various components of velocity? What is the normalized distribution function for x-components of velocity? What is the average value of v_x? Of v_y? Of v_z? What is the average of v_x^2? Of $mv_x^2/2$? Use the integral tables in Appendix B.

2-6. Suppose one cuts out a plot of $f(x)$ against x that has been made on cardboard of uniform density and balances the resulting piece of cardboard on a pin. If $f(x)$ is a probability density for x, to what will the x-coordinate of the balance point correspond?

2-7. Suppose the student population has joint probability density $f(w, V)$ for weight w of students and average beer consumed per day V given by the constant A in the ranges $100 \text{ lb} \leqslant w \leqslant 300$ lb, and $0 \leqslant V \leqslant 8$ pt. (*a*) What is the average "coefficient of inebriation," V/w? (*b*) Are weight and beer consumption correlated in this sample?

2-8. The gas molecules in a particular molecular beam have the following speed distribution function: $f(v)=a$ for $0\leqslant v\leqslant v_0$; $f(v)=2a$ for $v_0<v\leqslant 2v_0$; $f(v)=0$ for $v>2v_0$. Find the average speed and the most probable speed. What is the probability that a molecule will have speed between 0 and v_0?

* **2-9.** Consider the situation of Eq. 2-5 in which properties i and j are *not* mutually exclusive. Express the probability $P(i \text{ or } j)$ of finding at *least* property i or property j in terms of P_i, P_j, and the joint probability $P(i \text{ and } j)$.

2-10. Prove that Eq. 2-20 follows from Eq. 2-19.

2-11. If i and j are two properties of interest, the **conditional probability** of i under the condition $j, P(i|j)$, is defined by

$$P(i|j)=\frac{n_{i \text{ and } j}}{n_j}. \tag{2-94}$$

Note that the condition j means that all members of the ensemble that play any role must have j. Then the fraction of those that also have i is $P(i|j)$. If i and j are uncorrelated, express that fact in terms of $P(i|j)$ and discuss the result.

2-12. Conditional probabilities are defined in Prob. 2-11. The usual probabilities based on the complete ensemble are called **absolute probabilities**. Examples are

P_i and P_j. Express the joint probability P_{ij} in the general case *in two different ways* as the product of an absolute probability times a conditional probability.

2-13. Express $f_2(x,z)$ in terms of $f_4(w,x,y,z)$. Express $f_3(w,x,y)$ in terms of $f_4(w,x,y,z)$. Express the ensemble average of $h(w,z)$ in terms of an integral over $f_4(w,x,y,z)$. Express the same average as an integral over $f_2(w,z)$.

2-14. In the paragraph following Eq. 2-16, it is stated that P_{ij} contains more information than P_i and P_j do between them. Precisely what is the additional information? Under any circumstance would knowing P_i and P_j separately furnish as much information as knowing P_{ij}?

2-15. Derive Eqs. 2-46 through 2-56, based on the definition, Eq. 2-45, in analogy to what was done in Eqs. 2-5 to 2-43 for discrete variables.

2-16. Suppose each member of the ensemble is in a state labeled by the index i. Write the formal expression for the average of ln P, where P means the probability of a state. What range of values may $\overline{\ln P}$ take? What is $\overline{\ln P}$ when all members are in the same state?

3 THE ENSEMBLE IN STATISTICAL MECHANICS

THE PROBLEM OF STATISTICAL MECHANICS

Statistical mechanics addresses itself to the following problem: One has a **system,** that is, *that part of the physical world to which one directs his attention.* About this macroscopic system one knows certain information, usually as a result of measurements or observation. This information may be a minimum; for example, the volume, the mass of each constituent present, and the condition that the system is "in equilibrium at a given temperature," or "in equilibrium with a given energy." Or the information may be very extensive and detailed about a system evolving in some complicated nonequilibrium process. This we call the *initial information* about the system.

One also must have a *microscopic model* of the system, that is, a picture of the mechanical behavior of the particles that make up the system, be they molecules, atoms, ions, electrons, or photons.

The problem of statistical mechanics is to compute the probabilities of the various possible results of a variety of measurements one might choose to make on the system. The computation involves finding how the initial information and the known mechanical behavior of the constituent particles affect various other measurements. These other measurements might be made at the same time the initial information is taken. Or, the other measurements might be made at a later time, after the system has evolved for a while. Or, if the system is "in equilibrium," time is not even a factor in the problem.

One can compute only the *probabilities* of various results for these other measurements because they are underdetermined by the information at hand. However, for most *macroscopic* measurements the tremendous number of particles involved leads to the great simplification that the most probable result of a measurement proves to be overwhelmingly probable. As shown in Prob. 3-2, significant deviations from the mean or most probable value have vanishingly small probabilities. Thus once statistical mechanics has found the likely result of a measurement, that result may be safely predicted as the one that will be found for the system. The mean values of density and pressure in a gas, for example, are what we would *predict* for the system, even though the system *might* surprise us when we look at it by being in the midst of a significant *fluctuation*.* In principle, statistical mechanics can tell us the relative *probabilities* of *all possible* results of the measurement, as just discussed.

Thus, we have some information about a system we consider to be a mechanical object made up of constituent particles. The information falls far short of a complete mechanical specification of the state of the N particles (which in classical mechanics, e.g., would be the positions and velocities of all N particles). Yet mechanics is useful only when applied to particles in a completely specified mechanical state. (Imagine how useful classical mechanics would be if one knew only that the three balls were "someplace on the billiard table.") It is this very lack of detailed information that requires the addition of probability theory to the mechanics. Since we do not *know* what mechanical state the N particles are in, we say that each possible state has its own probability or probability density. This probability distribution for the various states is a function of the information that is known about the system. Thus *one constructs an ensemble to represent the system. Each member of the ensemble is a mental creation that is in a definite mechanical state. All members of the ensemble must be consistent with the initial information about the system.* Since those properties are known about the system, the probability of its not having those known

*See Andrews, Chap. 2, for a discussion of the problem posed by fluctuations to the very meaning of macroscopic properties.

properties is zero. All the members of the ensemble must reflect that information.

A clear distinction must be drawn between the system and a member of the ensemble. The system is the physical object about which we hope to make predictions. Members of the ensemble are only mental constructions on which the use of probability theory is based. Of course, we do not need to construct, even mentally, an infinite number of members with some 10^{23} particles in a fixed mechanical state.* We seek only the complete probability as a mathematical function of the necessary variables. Nevertheless, the ensemble and its members are useful to think about; they give a tangible feeling to the various probabilities, which might otherwise seem to be very abstract functions.

Choice of suitable variables to characterize the members of the ensemble depends on the definition of a completely specified mechanical state for the N particles. In turn, this depends on the kind of mechanics being employed in determining the microscopic picture of the system: classical mechanics, quantum mechanics,† or a mixture of both, which has acquired the name "semiclassical mechanics."

CLASSICAL MECHANICS‡

In classical mechanics, the state of an N-particle system is completely specified when the position $\mathbf{r}$ and momentum $\mathbf{p}$ (or velocity $\mathbf{v}$) of each particle are given.§ If the particles are polyatomic molecules, the position and momentum of each atom must be given. If the system is considered to be composed of N classical mechanical particles, the complete set of variables characterizing a member of the ensemble is the set of N position vectors $\mathbf{r}_1, \mathbf{r}_2, \mathbf{r}_3, \ldots, \mathbf{r}_N$ (this entire set is often noted as $\mathbf{r}^N$) that locates the particles and the set of N momentum vectors $\mathbf{p}_1, \mathbf{p}_2, \mathbf{p}_3, \ldots, \mathbf{p}_N$ (often noted $\mathbf{p}^N$), one for each particle. Since a vector is given by three numbers, say, its

*Statistical mechanics credits people with great *hypothetical* mental agility, but the *real* demands it makes on them are less unreasonable.

†In this book we never consider a consistent quantum mechanical treatment, which would be based on the density matrix. This subject is discussed by Tolman, Chap. IX. If required, its study should prove much simpler after reading this book.

‡Most of the larger references given in Appendix A, which cover either quantum mechanics or statistical mechanics, have chapters devoted to classical mechanics. The most thorough presentation is probably that of Tolman, Chaps. II and III, but the material is difficult.

§In this book, as in most of the statistical mechanical literature, vectors are denoted by boldface type. The origin of position vectors is arbitrary and can be taken to be one corner of the box holding the system. The origin of momentum or velocity vectors is chosen naturally by the momentum or velocity of the system as a whole.

x, y, and z components, there are $6N$ continuous variables characterizing each member of a classical ensemble.

The momentum has been indicated here, rather than velocity, even though momenta and velocities are often simply related. This has been done because classical statistical mechanics is in general much simpler when the variables used are the positions of the particles and the momenta appropriate to these positions. Throughout physics, momenta appear as more fundamental mechanical variables than velocities.

The $3N$ position components plus the $3N$ appropriate momentum components, which together form the complete set of variables for the classical ensemble, have been given the name of the **phase space** of the classical N-particle system. Just as a point in ordinary space is the specification of the three coordinates that locate it, a point in phase space is the specification of the $6N$ coordinates $(\mathbf{r}^N, \mathbf{p}^N)$ that locate it. Thus ordinary Cartesian space is three-dimensional and phase space is $6N$-dimensional. Each member of the classical ensemble is completely characterized by a single point in phase space. Sometimes the $3N$ position components $\mathbf{r}^N$ are referred to separately as the **configuration space** of the problem and the $3N$ momentum components $\mathbf{p}^N$ as the **momentum space.** This makes the complete phase space the sum of configuration space plus momentum space.

Thus classical statistical mechanics has the following conceptual structure: we know a certain amount about a system consisting of N particles whose mutual mechanical interactions are presumed to be known. We construct mentally an ensemble to represent the system and what we know about it. Each member of the ensemble has the position and momentum of each particle specified, with different specifications for the different members, yet each specification is consistent with the initial information. The result is a complete probability amplitude $f_N(\mathbf{r}^N, \mathbf{p}^N)$ for the $6N$ variables that describe the classical state of a member of the ensemble. The f_N is defined as follows: $f_N(\mathbf{r}_1, \mathbf{r}_2, \mathbf{r}_3, \ldots, \mathbf{r}_N, \mathbf{p}_1, \mathbf{p}_2, \mathbf{p}_3, \ldots, \mathbf{p}_N)\,d\mathbf{r}_1\,d\mathbf{r}_2\,d\mathbf{r}_3 \cdots d\mathbf{r}_N\,d\mathbf{p}_1\,d\mathbf{p}_2 \cdots d\mathbf{p}_N \equiv f_N(\mathbf{r}^N, \mathbf{p}^N)\,d\mathbf{r}^N\,d\mathbf{p}^N$ is the joint probability that particle 1 has its x-component of position between x_1 and $x_1 + dx_1$, its y-component of position between y_1 and $y_1 + dy_1$, its z-component of position between z_1 and $z_1 + dz_1$ (i.e., its position lies in the volume element $d\mathbf{r}_1$ centered about the point $\mathbf{r}_1$), its x-component of momentum between p_{x1} and $p_{x1} + dp_{x1}$, its y-component of momentum between p_{y1} and $p_{y1} + dp_{y1}$, its z-component of momentum between p_{z1} and $p_{z1} + dp_{z1}$ (i.e., its momentum lies in the range $d\mathbf{p}_1$ centered about $\mathbf{p}_1$), that particle 2 has its position in $d\mathbf{r}_2$ centered about $\mathbf{r}_2$ and its momentum in $d\mathbf{p}_2$ about $\mathbf{p}_2$, and that particle 3 has $\cdots$ etc. for all N particles. The functional dependence of f_N on its $6N$ variables is determined in a way that reflects the initial information. The ensemble is then used to generate all the various probabilities of interest, either by using f_N itself or else, more commonly, by using reduced distribution functions like f_1 and f_2 obtained from f_N by integrating over the uninteresting variables in accordance with Eqs. 2-49 and 2-50.

Methods of constructing the function $f_N(\mathbf{r}^N, \mathbf{p}^N)$ for equilibrium systems are discussed in Chap. 17. Methods of constructing it for simple nonequilibrium systems have been discussed by F. C. Andrews, *J. Chem. Phys.*, **47**, 3161 (1967).

QUANTUM MECHANICS*

Of course, it is not classical mechanics but quantum mechanics that truly describes the state of matter. In quantum mechanics the N-particle stationary system contained in a finite volume may exist in any one of an enormous number of discrete states, found in principle by solving the **Schrödinger equation** written symbolically as

$$\mathcal{H}_N \Psi_i(\mathbf{r}^N) = E_i \Psi_i(\mathbf{r}^N). \tag{3-1}$$

Here Ψ_i is the **wave function** *for the entire N-particle system* in state i, and it is a function of the $3N$ position coordinates. (Note: Many students are accustomed to using an equation of the form of Eq. 3-1 only for simple one- or two-particle systems like the particle in a box, the hydrogen atom, the harmonic oscillator, or the rigid rotor. However, Eq. 3-1 is the correct starting point for any stationary system, however complicated, and we are using it here for the complete N-particle system.) The **Hamiltonian operator** $\mathcal{H}_N$ is obtained through the usual prescription of writing the total energy of the N particles in terms of $\mathbf{r}^N$ and $\mathbf{p}^N$, then replacing $\mathbf{p}_j$ by $-i\hbar\nabla_j$, where ∇_j is the gradient operator for the coordinates of particle j. The **energy eigenvalue** E_i is the total energy the N-particle system would have if it were in the ith quantum state. The solution of Eq. 3-1, with appropriate conditions on Ψ at the walls of the system and with due attention to the failure of the interchange of like particles to produce a new quantum state, leads to the identification of an enormous number of discrete states in which the system can exist. It may be in any one of the allowed states, say state i; it will then have the energy E_i associated with that state and it will be described by the wave function Ψ_i. But the states arising from the

*There are many satisfactory books on quantum mechanics, and a few are listed in Appendix A. Also, most of the larger books cited there that deal with statistical mechanics contain chapters devoted to quantum mechanics.

solution of Eq. 3-1 represent an exhaustive specification of the possibilities for the system. It must be in one or another of them.

Since the allowed N-particle quantum states are discrete, they can be ordered in some way, usually in order of increasing energy, and a number can be assigned to each. This single number i completely specifies the state as that represented by the wave function Ψ_i and whose energy is E_i. Thus in quantum mechanics each member of the ensemble is characterized by the single, discrete number that labels its state, with which is associated the appropriate wave function and energy eigenvalue.

Thus quantum statistical mechanics has the following conceptual structure. Initially, we know a certain amount about a system consisting of N particles whose mutual interactions are presumed known. We solve the Schrödinger equation to determine the set of allowed quantum states the system could be in. We construct mentally an ensemble to represent what we know about the system. Each member of the ensemble is in one of the quantum states permitted the system; only states consistent with the initial information are contained in the ensemble. The result is a probability P_i, which is a function of the discrete number i that labels the N-particle quantum states. The initial weighting of the different states that determine P_i is done in a way that reflects the initial information. The ensemble is then used to generate all the various probabilities of interest.

Methods of constructing the function P_i for equilibrium systems and of using it to calculate various averages are discussed in the remainder of Part 1 of this book.

Since nature is in fact described by quantum mechanics, one might ask why classical mechanics with its $6N$ different continuous variables is ever used in statistical mechanics. The reasons are several. Quantum mechanics, having a single discrete parameter, is simple in principle, but except for ideal gases, the job of finding the allowed quantum states for an N-particle system is usually extremely difficult. In classical mechanics, that part of the problem at least is solved. Also, in classical mechanics it is easier to visualize N particles, each having its position and momentum, than it is an N-particle quantum state. In some cases the quantum and classical pictures become identical, and these occasions help one learn how to choose the appropriate description to simplify the problem at hand.

SEMICLASSICAL MECHANICS

One last description of the microscopic picture is often useful. In many cases, the positions and momenta of the particles can be described very adequately by classical mechanics. However, the particles themselves sometimes have complicated structure (e.g., polyatomic molecules). The internal state of a molecule may almost never be described adequately by classical mechanics. A quantum description is necessary. In this case, we describe the centers of mass of the molecules by classical positions and momenta and use quantum mechanics to describe the internal states of the molecules. Such a mixed or hybrid description has commonly been called *semiclassical.*

The variables needed to describe a member of a semiclassical ensemble are as follows: the set of continuous positions $\mathbf{r}^N$ and momenta $\mathbf{p}^N$ for the centers of mass of the N particles, and the set of discrete quantum numbers needed to fix the internal states of each of the N molecules. This set might include rotational, vibrational, electronic, and even nuclear quantum numbers. If ν different quantum numbers describe the internal state of each molecule, the semiclassical ensemble is completely described by $6N$ continuous and νN discrete variables.

The purely classical description is adequate only when one has no interest in the internal states of the molecules (e.g., in the study of a monatomic fluid like a noble gas). If all the atoms are in their lowest electronic state, the fact that the atoms have a structure of their own can be neglected. As the temperature is increased and electronic excitation becomes more probable, the structure begins to play an active role in the properties. Then even gases like argon must be treated semiclassically. In the latter parts of the book, we have occasion to use all three descriptions.

DISCUSSION

Just how valid is the statistical mechanical approach? It is likely that it represents the ultimate in "scientific truth." All measurements on systems yield partial information. If one were able to look at a system and learn the exact position, momentum, and internal state of each particle, the ensemble would have to reflect that information. Each member of the ensemble would then have to be identical. An ensemble consisting of just one member would be adequate, and that member would be identical to the system. In that case it would be silly to introduce the concept of probability into the many-particle mechanics used to study the system. One would simply confront a problem in *mechanics;* there would be no need for *statistical* mechanics.

However, complete information is never obtained. With the partial information obtainable, one still wishes to predict as accurately as possible the results of measurements. Statistical mechanics can give this prediction and give it correctly—it consists in the distribution of probabilities of various measurements on a system, which is, naturally, a function of the information previously known about the system.

In the long run, as with any theory, the success of statistical mechanics rests on its usefulness in practice. It has proved very useful whenever its mathematical difficulties have not been excessive. Its usefulness in the future in more complex cases will depend on whether it can be visualized more and more simply, thus easing the mathematical difficulties. The development of large computers certainly has helped.

PROBLEMS

3-1. Explain the meanings of the following sets of symbols: $\mathbf{r}_i$; $\mathbf{p}_i$; $\mathbf{r}^N$; $\mathbf{p}^N$; $f(\mathbf{r}^N, \mathbf{p}^N)$; $d\mathbf{r}_i$; $d\mathbf{p}_i$; $d\mathbf{r}^N\, d\mathbf{p}^N$.

3-2. The average kinetic energy of a mole of gas molecules is $\frac{3}{2}RT$. Suppose you were measuring the energy of a gas at 1 atm and 25°C with great precision, say to ± 0.001 cal, and a large cluster of molecules suddenly behaved peculiarly by acquiring twice their average energy. How many molecules would have to be involved in order for this fluctuation to be barely noticeable? What does this imply about wild fluctuations involving 10^5 or even 10^{10} particles?

4 EQUILIBRIUM AND THE ISOLATED SYSTEM

THE EQUILIBRIUM CONDITION

The second law of thermodynamics states that any system isolated from the rest of the world will run down and reach equilibrium.* This means that its macroscopic properties will reach time-independent values. This book is restricted to studying the statistical mechanics of equilibrium systems; nonequilibrium statistical mechanics is marked by great mathematical difficulty and a large number of still-unresolved problems.

*There are other statements of the second law, but this one is the most intuitive. All are discussed at considerable length, along with the concept of equilibrium, in Andrews, Chaps. 8–11.

From the viewpoint of statistical mechanics, *a system, regardless of the state in which it started, is in a condition of* **equilibrium** *when the information one has about it has reached a time-independent minimum.* The information is usually one of the following three sets:

1. The system consists of N particles of known type or types contained in a volume V. It is *completely isolated* from the rest of the world by fixed insulating walls. This means, from the first law of thermodynamics, that its energy has remained constant from the moment the system was constructed, and the energy of the resulting equilibrium system is that known, constant value. The isolated equilibrium system is therefore completely described by N, V, and E. This condition is analyzed in this chapter; the ensemble used to represent this information is called *microcanonical.*

2. The system consists of N particles of known type or types contained in a volume V. It is in *thermal contact* with a very large heat reservoir, which is simply a huge equilibrium system, much larger than "the system," which is characterized by the temperature T. This means that although energy fluctuations are possible for the system, since it can exchange energy with the heat reservoir, its temperature at equilibrium will by definition be that of the reservoir, as is consistent with the zeroth law of thermodynamics. The equilibrium system in thermal contact with a heat reservoir is therefore completely described by N, V, and T. This condition is analyzed in Chap. 5; the ensemble used to represent this information is called *canonical*.

3. The system consists of the material contained in a volume V. This material is in *thermal contact* with a large heat reservoir characterized by the temperature T. The system is also *open to the exchange of each type of matter* with a particle bath or reservoir for each type. These are simply huge equilibrium systems containing particles of the appropriate type, characterized in each case by the chemical potential μ for the particular species, and in contact with the system through a wall of the system which is permeable to that kind of particle. A simple example of such a system would be to define it as a particular cubic centimeter of air in the laboratory. The gas in that cubic centimeter is in thermal equilibrium with the air in the rest of the room, which serves as a heat bath for it and is characterized by the temperature T. Also, nitrogen, oxygen, argon, and other gases, can pass through the walls to enter or leave the system. The chemical potentials for N_2, O_2, Ar, and so on, are established in the particle bath, which in this example doubles as the heat bath. The equilibrium system in thermal contact with a heat reservoir and open to the exchange of matter with a particle bath is therefore completely described

by μ, V, and T. This condition is analyzed in Chap. 16; the ensemble used to represent this information is called *grand canonical.*

In all these cases and others treated by different equilibrium ensembles, the system is known to be "in equilibrium." One associates both time independence and a certain constancy in properties with the equilibrium condition. However, on examining more and more closely any macroscopic "equilibrium" system, one learns much specific information about that system. Local variables, such as the density and pressure, are found to fluctuate from place to place in the system. These fluctuations, caused by the motion of the particles in the system, are constantly arising and being dissipated. Thus if one knows too much about a system that would otherwise be considered "in equilibrium," he destroys the constancy of properties associated with the equilibrium condition.

The ensemble that correctly represents the equilibrium condition must simultaneously represent all possible fluctuations, each weighted according to its probability. Constancy in properties is not required of a system known to be "in equilibrium." The extent to which constancy is expected should be calculable from the probability function, however.

Time independence, nevertheless, is certainly part of the equilibrium condition. The time independence rests *not* in the properties of the physical system, but in the state of knowledge one has about the system. With condition 1 above, the investigator knows that the volume is constant, the number and type of particles are constant, and the energy of the system is constant. Furthermore, his own state of ignorance about any other features of the system is constant. Thus based on that information there must be a certain constant set of *predictions* about the system, that depend only on the N, V, and E, and the fact that the system is "in equilibrium." The system may have got into its equilibrium condition from any of a number of previous conditions, but it has sat around long enough to be now "at equilibrium." Information about its former condition is completely lost or meaningless, and it remains lost unless more detailed observations are made.

THE MICROCANONICAL ENSEMBLE

We consider an equilibrium system in condition 1 above; it is isolated, and its condition is specified by the values of N, V, and E. We now confront the problem of building that information into an ensemble to represent the equilibrium system.

The first task is always to solve the Schrödinger equation, 3-1, for the set of particles N in the volume V. Knowledge of N and V is sufficient to formulate Eq. 3-1, and in principle to solve it. The result is a set of quantum states for the N-particle system, labeled by the number i, each state with its own energy eigenvalue E_i. This result already reflects the values of N and V. We now want to find how P_i depends on i.

The only additional knowledge possessed is that the system is isolated with fixed energy E_{system}. The first, obvious step in determining how P_i depends on i is to assign zero probability to all states i unless E_i equals the measured system energy E_{system}. All members of the ensemble must be in quantum states with $E_i = E_{\text{system}}$. Beyond this we run into trouble, however, because for macroscopic systems there are usually an *enormous* number of different quantum states i characterized by $E_i = E_{\text{system}}$. How do we weight each of these states in the equilibrium ensemble?

The number of different quantum states which have a given energy is called the **degeneracy** *of the energy level.* (Note: we consistently use the word *level* to refer to the value of the *energy* for one or more states; thus energy levels have degeneracies; quantum states do not.) As the energy of a macroscopic system increases, the degeneracies of the different energy levels increase drastically, and we confront the problem of giving proper weight in the ensemble to each degenerate state of the energy level E_{system}. Each such state represents a rich detail of information about the properties the system would have if it were in that state. Yet no such detailed information is available, and each such state must be assigned its probability. Since the only information about the dynamics of the system that we have is the system energy, we place the following requirement on the probability P_i: *the only dynamical feature on which the probability of a state may depend is the energy of the state.* The probability P_i of the ith N-particle quantum state is a function of E_i only:

$$\boxed{P_i = P_i(E_i \text{ only}).} \tag{4-1}$$

This requirement or hypothesis calls for some discussion, since it is the basis of all of equilibrium statistical mechanics. Suppose the probabilities of the allowed quantum states were assigned as functions of some quantity in addition to the energy—for example, the number of particles that a system in that state would have in a particular cubic centimeter of the system. That would mean that for some reason we were biasing the probability distribution for or against certain numbers of particles in that cubic centimeter. However, without having measured the number of particles there, we have no right to bias the distribution in that way. And if we had made the measurement, we would have too much information for an "equilibrium" system; moreover, the time independence of our predictive ability would be destroyed, as discussed above. The concept of information is elaborated in Chap. 7.

Thus the basic hypothesis of Eq. 4-1 seems to have been forced on us. Anything other than the energy on which the probabilities might be allowed to depend would not have been part of the initial information available under these circumstances. Such additional information would have been spurious, thus it has no place in the ensemble. Equation 4-1 may be viewed either as the basic assumption of equilibrium statistical mechanics, or as a direct consequence of the aims of statistical mechanics.

Examination of Eq. 4-1 shows that *all quantum states with the same energy* $(E_i = E_j = E_k = E)$ *have the same probability* $(P_i = P_j = P_k = P(E))$. The probabilities of all degenerate quantum states with the same energy are equal. This solves the problem of representing the isolated equilibrium system. If the degeneracy of the energy level E_{system} is W, those W different states are the only ones which are represented in the ensemble. Each such state has equal probability, whose value can be found following Eqs. 2-8 to 2-11, with $f(i)$ equal to a constant:

$$P_i = a, \qquad (E_i = E_{\text{system}}) \tag{4-2}$$

$$\sum_{i=1}^{W} P_i = \sum_{i=1}^{W} a = Wa = 1, \tag{4-3}$$

$$a = 1/W. \tag{4-4}$$

An ensemble constructed in this way was named by J. Willard Gibbs a **microcanonical ensemble,** which generates the so-called **microcanonical** probability **distribution,** given by

$$P_i = \begin{Bmatrix} 1/W & (E_i = E_{\text{system}}) \\ 0 & (E_i \neq E_{\text{system}}) \end{Bmatrix}. \tag{4-5}$$

Only the degenerate states with energy E_{system} have nonvanishing probabilities, the same constant value for all such states. Gibbs did not say why he chose this name, but the term has stuck. The "micro" prefix may be thought of as connoting the *micro*scopically sharp energy requirement.* The word "canonical" connotes the *general acceptance* of the ensemble

*The measured system energy must always be slightly uncertain, due to the Heisenberg uncertainty principle. Broadening slightly the sharp energy requirement of this chapter does not change the results in any important way.

described by Eq. 4-5 as representative of the isolated equilibrium system. More likely, a microcanonical ensemble was conceived by Gibbs as consisting of a microscopically thin slice taken from a canonical ensemble, as described in the next chapter.

Figure 4-1 is a plot of the microcanonical distribution, showing P_i as a function of E_i. The value W is actually W_{system}, the degeneracy of the energy level E_{system}. Degeneracies of energy levels increase with energy in a manner similar to that represented in Fig. 4-2. Thus the greater the measured system energy, the more quantum states are available to the system, and the smaller the probability of any one state.

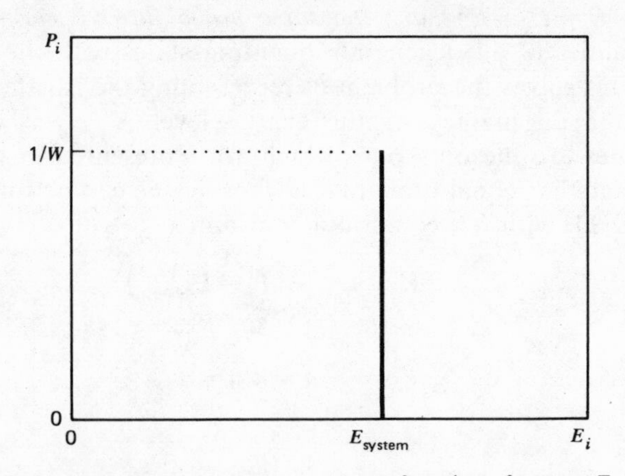

Fig. 4-1. Probability of N-particle quantum state i as function of energy E_i of state i for an isolated equilibrium system. The microcanonical distribution.

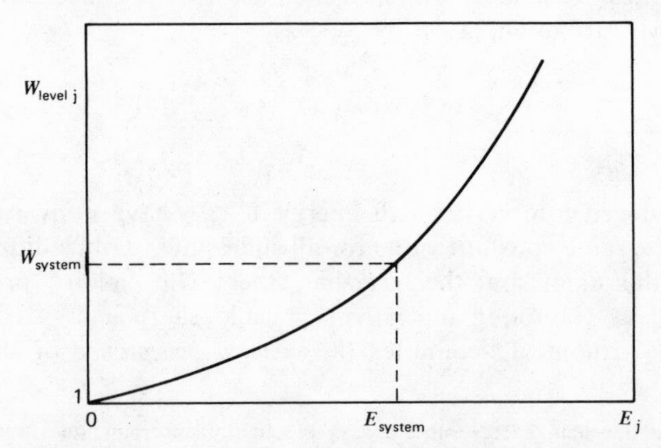

Fig. 4-2. Typical plot of degeneracy W_j of the jth energy level as a function of energy E_j. There are W_j different N-particle quantum states having the same energy E_j. This plot drastically underrepresents both the slope and the rate of increase of the slope of such curves.

We conclude this chapter by raising the question of whether an ensemble constructed in some way according to Eq. 4-1 to represent an equilibrium system would indeed remain independent of time as each member evolved. The answer is that it would, and the proof is based on a rather general conclusion from quantum mechanics called the *principle of detailed balancing.** According to this principle, the probability that a member of the ensemble in state A will go into state B is the same as that of one in state B going into state A. The transition is only allowed at all if energy is conserved, such that the energies of A and B are equal. Therefore, if an ensemble is constructed to have the same number of members in both A and B, on average the same number per unit time will go from A to B as from B to A. Thus the structure of the ensemble remains constant in time.

PROBLEMS

4-1. Summarize the necessary and sufficient conditions for the validity of the microcanonical distribution of probabilities of N-particle quantum states, as given in Eq. 4-5.

4-2. As presented here, Eq. 4-1 might seem to be a reasonable basis for statistical mechanics. What is the *real* justification for its use?

4-3. Conventional coin-tossing probabilities are familiar examples of the microcanonical ensemble. Impossible results (e.g., coin stays up in the air, or coin lands and remains on edge) are assigned zero probability. Each possible state (head or tail) of a single coin is given equal probability if no information is available about whether the coin is biased. If two coins are tossed, how much less likely are 2 heads (or 2 tails) than 1 of each? If four coins are tossed, how much less likely are 4 heads (or 4 tails) than 2 of each? If six coins are tossed, how much less likely are 6 heads (or 6 tails) than 3 of each?

4-4. Choice of a zero of energy is always arbitrary; the scientist ordinarily chooses it simply for convenience in working with the problem at hand. How does choice of zero of energy affect the microcanonical distribution?

*For example, see Tolman, p. 521.

5 SYSTEM IN EQUILIBRIUM WITH A HEAT BATH

THE NATURE OF A HEAT BATH

One is less likely to encounter an isolated equilibrium system than a system in thermal equilibrium with a **heat bath**. If the system and heat bath are in equilibrium with each other through any diathermal wall, they will "be at the same temperature," according to the zeroth law of thermodynamics.* Temperature is not, however, a mechanical quantity like the energy; therefore, we must give attention to how it arises in statistical mechanics. Our picture of a heat bath is of an infinitely large system at equilibrium; its relationship to the system is schematized in Fig. 5-1. Thus the bath is capable of giving up or receiving any finite amount of energy through the diathermal wall without itself changing appreciably. This means that no restriction whatever is placed on the value of the energy of the system; E_{system} will fluctuate about an average value $\overline{E}$ as the system interacts randomly with the heat bath. One does not *know* the system's energy; it might have any value.

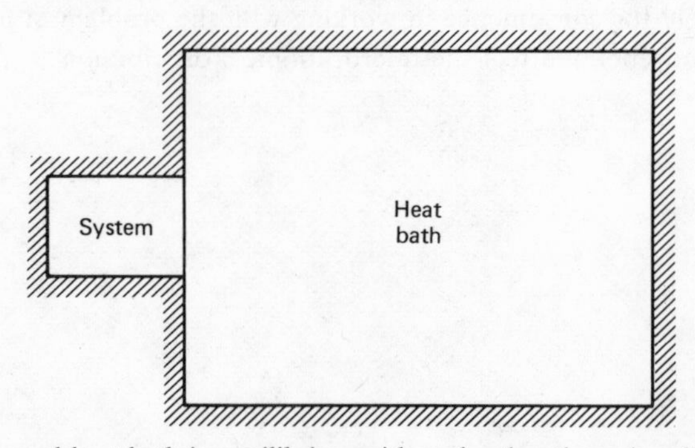

Fig. 5-1. System and heat bath in equilibrium with each other through a diathermal wall.

*Andrews, pp. 40–44.

The problem of establishing how the probability P_i of the ith N-particle quantum state (where the system contains N particles) depends on the value of energy E_i is therefore more complicated in this situation than it was for the isolated system of Chap. 4. We start with the basic postulate of Eq. 4-1, namely, that the only dynamical feature on which the probability of a state may depend is the energy of the state:

$$P_i = P_i(E_i \text{ only}). \tag{5-1}$$

The rationalization of Eq. 5-1 for systems in equilibrium with heat baths is not quite so convincing as it was for the isolated systems of Eq. 4-1. Once again, however, it involves recognizing how arbitrary it would be to insert some other feature of the state into the parentheses of Eq. 5-1, when no such other feature was observed. And again, we note that if it had been observed, the information gained would have been excessive for a system truly in equilibrium. In the long run, the justification of Eq. 5-1 rests in the continuing agreement between experimental results and statistical mechanical predictions made from probability distributions employing it.

Determination of the functional dependence of P_i on E_i proceeds by recognizing that the energy of the N-particle system in quantum state i can almost always be interpreted meaningfully with great precision as the *sum* of a host of small energy contributions. These may be, for example, the kinetic energies of individual particles, the potential energies of the particles relative to external fields, the internal energies of the various particles (e.g., the nuclear energies of atoms, the electronic energies of atoms and molecules, the rotational and vibrational energies of molecules), the intermolecular potential energies between pairs or triples of particles, correction terms for the oversimplification of this analysis, and for solids the energies of the various normal modes of vibration, and perhaps others. This fact of summation may be written

$$E_i = \epsilon_1 + \epsilon_2 + \epsilon_3 + \cdots = \sum_j \epsilon_j, \tag{5-2}$$

where the ϵ_j are an enormous number (at least of order N) of small contributions, each capable of more or less intuitive understanding as being of a relatively independent or localized "part" of the system.

Combining Eq. 5-2 and 5-1 yields

$$P_i = P_i(\epsilon_1 + \epsilon_2 + \epsilon_3 + \cdots). \tag{5-3}$$

The probability P_i of the ith N-particle quantum state is being viewed as the joint probability that each "part" j of the system has just the right energy ϵ_j to permit all the parts to combine to yield the N-particle state i. We now ask what might be present in the system to correlate explicitly the energies of these various "parts" and prevent the factoring:

$$P_i(\epsilon_1+\epsilon_2+\epsilon_3+\cdots)=p_1(\epsilon_1)p_2(\epsilon_2)p_3(\epsilon_3)\cdots. \tag{5-4}$$

For example, what might cause the amount of kinetic energy of a particle in one region of the system to be correlated to the amount of rotational energy of another particle, off in a distant region of the system? Such a correlation would imply that the probability distribution for kinetic energy of the first particle would change with the knowledge of whether the second particle had large or small rotational energy. We have already seen such a correlation in Chapter 4: recall that the microcanonical distribution Eq. 4-5 does not factor, the correlation having been caused by the *requirement on the total energy,* that $\epsilon_1+\epsilon_2+\epsilon_3+\cdots$ had to equal E_{system}. Thus if the second particle had a huge amount of rotational energy, there would be less energy available for the first particle, and its distribution for kinetic energy would consequently be skewed to favor lower values of ϵ. But there is no such requirement on the total energy when the system is in equilibrium with a heat bath, rather than being isolated. All total energies are possible, and if the second particle has a huge rotational energy, the first can still have as much kinetic energy as happenstance gives it, because the heat bath is an infinite source or sink of energy.

Since a requirement on the total system energy is the only cause we can imagine that might prevent the factoring of Eq. 5-4, and since such a requirement does not exist because of the heat bath, we accept the factoring of P_i as in Eq. 5-4, and we now explore the consequences of that factoring on Eq. 5-3. Taking the logarithm of both sides of Eq. 5-4 yields

$$\ln P_i(\epsilon_1+\epsilon_2+\epsilon_3+\cdots)=\ln p_1(\epsilon_1)+\ln p_2(\epsilon_2)+\ln p_3(\epsilon_3)+\cdots. \tag{5-5}$$

The left-hand side of Eq. 5-5 is a function of the sum of the ϵ's, since the logarithm of a function of the sum is itself another function of the sum.* The left-hand side can therefore be expanded in a power series in its argument (as $f(x)=a_0+xa_1+x^2a_2+x^3a_3+\cdots$ for any arbitrary function f

*The following alternative derivation of Eq. 5-7 from 5-5 is preferred by some. If Eq. 5-5 is differentiated with respect to one of the ϵ's, say ϵ_j, the right-hand side is $\partial \ln p_j(\epsilon_j)/\partial\epsilon_j$, and the left-hand side is $\partial \ln P_i(E_i)/\partial\epsilon_j=[d\ln P_i(E_i)/dE_i][\partial E_i/\partial\epsilon_j]=d\ln P_i(E_i)/dE_i$, where we have used the chain rule and recognized $\partial E_i/\partial\epsilon_j$ to be unity. Now how can $d\ln P_i(E_i)/dE_i$ (which is a function of the *sum* of all the ϵ's) equal $d\ln p_j(\epsilon_j)/d\epsilon_j$ (which is a function of only one ϵ)? Only if both are constants. Thus $d\ln P_i(E_i)/dE_i=$constant, integration of which yields $\ln P_i=a_iE_i+a_0$, which is precisely Eq. 5-7.

of x that is well behaved at $x=0$), which in this case is the sum of ϵ's:

$$\ln P_i(\epsilon_1+\epsilon_2+\epsilon_3+\cdots)=a_0+(\epsilon_1+\epsilon_2+\epsilon_3+\cdots)a_1$$

$$+(\epsilon_1+\epsilon_2+\epsilon_3+\cdots)^2a_2+\cdots, \qquad (5\text{-}6)$$

where the a's are arbitrary quantities, independent of the ϵ's, whose values must be established by additional specific information. The physical content of Eq. 5-6 is just that of Eq. 5-3. Now, however, we note from Eq. 5-5 that for P_i to factor, $\ln P_i$ must equal a sum of functions of each separate ϵ, with no cross terms that couple two or more different ϵ's. That can be the case only if a_2 and all higher a's are identically zero. If a_2, for example, were not zero, Eq. 5-6 would generate terms like $\epsilon_1\epsilon_2a_2$ which are not present in the right-hand side of Eq. 5-5 and could not be canceled by any number of such higher order terms as $\epsilon_1\epsilon_2^2a_3$. Thus the requirement that $P_i(E_i)$ must factor leads rigorously to the conclusion that

$$\ln P_i=a_0+(\epsilon_1+\epsilon_2+\epsilon_3+\cdots)a_1, \qquad (5\text{-}7)$$

$$P_i=e^{a_0+(\epsilon_1+\epsilon_2+\epsilon_3+\cdots)a_1}=e^{a_0}e^{a_1E_i}, \qquad (5\text{-}8)$$

$$\boxed{P_i=ae^{-\beta E_i}.} \qquad (5\text{-}9)$$

In Eq. 5-9 we simply defined two new constants:

$$a\equiv e^{a_0} \quad \text{and} \quad \beta\equiv -a_1. \qquad (5\text{-}10)$$

Gibbs calls the probability distribution represented by Eq. 5-9 the **canonical distribution,** naming ensembles constructed in this way **canonical ensembles.**

The parameter a is found immediately by normalizing Eq. 5-9 to unity:

$$\sum_i P_i=1=a\sum_i e^{-\beta E_i} \qquad (5\text{-}11)$$

$$a=\frac{1}{\sum_i e^{-\beta E_i}}\equiv\frac{1}{Q} \qquad (5\text{-}12)$$

$$P_i=\frac{e^{-\beta E_i}}{\sum_i e^{-\beta E_i}}\equiv\frac{e^{-\beta E_i}}{Q}. \qquad (5\text{-}13)$$

In Eq. 5-12 the sum over all states of $e^{-\beta E_{\text{state}}}$ is given the symbol Q and is called the **partition function** (see discussion of Eq. 8-6). The result, Eq. 5-13, is the normalized expression for the probability of an N-particle quantum state for a system in equilibrium with a heat bath. The system's volume and number and kind of particles were reflected in the solution of the Schrödinger equation, which yielded the set of quantum states and their energies E_i. The only other constraint on the system is its thermal equilibrium with the heat bath. Since the zeroth law says that the only distinguishing feature of the heat bath is its temperature, we might expect the only undetermined feature of Eq. 5-13—namely, β—to turn out to depend only on T. This indeed is demonstrated in the next chapter.

A plot of P_i versus E_i for the canonical distribution, comparable to that of Fig. 4-1 for the microcanonical, appears in Fig. 5-2. The difference between the two is the removal of the microscopically sharp energy requirement. The system may have any energy; the higher the energy of a quantum state, the less likely the system is to be in that state. However, all N-particle quantum states with the same energy have the same probability. We might think of a canonical ensemble as constructed from a large number of microcanonical ensembles, one for each different possible value of E, weighted according to Eq. 5-9.

Knowing P_i, it is easy to calculate the ensemble average of the energy $\overline{E}$ of the N-particle system using Eq. 2-28: $\overline{E} = \Sigma_i P_i E_i$:

$$\boxed{\overline{E} = \frac{\sum_i e^{-\beta E_i} E_i}{\sum_i e^{-\beta E_i}}} \tag{5-14}$$

or

$$\boxed{\overline{E} = -\left(\frac{\partial \ln Q}{\partial \beta}\right)_{V,N}}. \tag{5-15}$$

Equation 5-14, the first expression for $\overline{E}$, follows directly on use of Eq. 5-13 for P_i. Proof that Eq. 5-15 is an equivalent way of writing Eq. 5-14 is obtained by working backward from Eq. 5-15:

$$-\frac{\partial \ln Q}{\partial \beta} = -\frac{1}{Q}\frac{\partial Q}{\partial \beta} = -\frac{1}{\sum_i e^{-\beta E_i}} \frac{\partial \Sigma_j e^{-\beta E_j}}{\partial \beta} \tag{5-16}$$

$$= -\frac{1}{\Sigma_i e^{-\beta E_i}} \sum_j -E_j e^{-\beta E_j}, \tag{5-17}$$

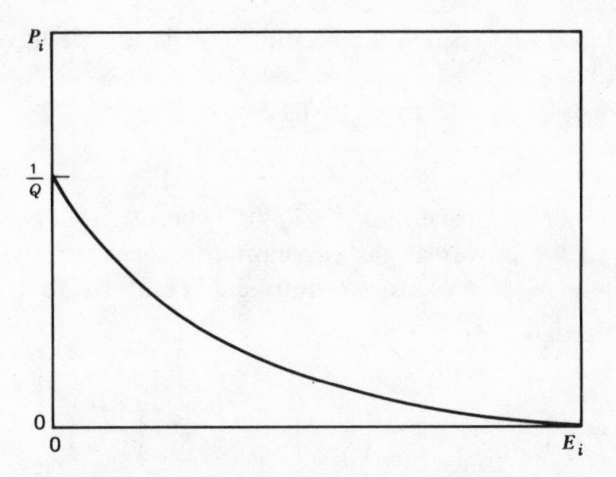

Fig. 5-2. Probability of N-particle quantum state i as function of energy E_i of state i for a system in equilibrium with a heat bath. The canonical distribution.

which is the same as Eq. 5-14. Thus for $\bar{E}$ we have two equivalent expressions, Eqs. 5-14 and 5-15, in terms of the set of N-particle quantum states, their energies E_i, and the parameter β.

In thermodynamics, whether a system is isolated or in thermal equilibrium with a heat bath is immaterial to the final equilibrium condition. Both systems have a definite thermodynamic energy, and even the latter system has only negligible energy fluctuations about this thermodynamic energy. If statistical mechanics is to agree with macroscopic observations, one must be able to predict that the energy of the system *will be* $\bar{E}$ and that significant fluctuations about $\bar{E}$ are highly unlikely. To show this, let us find the mean square deviation of the energy. From Eq. 2-40 we know

$$\overline{\delta E^2} = \overline{E^2} - \bar{E}^2. \tag{5-18}$$

Differentiating Eq. 5-14 with respect to β yields

$$\frac{\partial \bar{E}}{\partial \beta} = \frac{-\Sigma_i e^{-\beta E_i} E_i^2}{\Sigma_i e^{-\beta E_i}} - \frac{\Sigma_i e^{-\beta E_i} E_i \Sigma_j - e^{-\beta E_j} E_j}{\left(\Sigma_i e^{-\beta E_i}\right)^2}. \tag{5-19}$$

where we have followed the formula

$$\frac{\partial}{\partial \beta}\frac{a}{b} = \frac{(\partial a/\partial \beta)}{b} - \frac{a(\partial b/\partial \beta)}{b^2}. \tag{5-20}$$

The first term on the right-hand side of Eq. 5-19 is recognized immediately from Eq. 2-28 as $-\overline{E^2}$. Since the second term is $\overline{E}^2$, we have proved that

$$\overline{\delta E^2} = -\left(\frac{\partial \overline{E}}{\partial \beta}\right)_{V,N}. \tag{5-21}$$

In the next chapter we prove that $\beta = 1/kT$, where k is a constant found to be analogous to the universal gas constant R, except for a single particle: $k = R/N_0$, where N_0 is Avogadro's number. We utilize that result here to find the magnitude of $\overline{\delta E^2}$:

$$\overline{\delta E^2} = -\left(\frac{\partial \overline{E}}{\partial \beta}\right)_{V,N} = -\left(\frac{\partial \overline{E}}{\partial T}\right)_{V,N}\frac{dT}{d\beta} = -\left(\frac{\partial \overline{E}}{\partial T}\right)_{V,N} \div \frac{d\beta}{dT}, \tag{5-22}$$

$$\overline{\delta E^2} = kT^2\overline{C}_V, \tag{5-23}$$

where $(\partial \overline{E}/\partial T)_V$ was recognized as the ensemble average of the heat capacity at constant volume, by the definition of C_V.* The question of whether a quantity like $\overline{\delta E^2}$ is large or small is meaningless until one establishes "large or small *compared to what?*" The commonest procedure is to find the ratio of the root mean square deviation to the average value of the quantity under consideration:

$$\frac{\sqrt{\overline{\delta E^2}}}{\overline{E}} = \frac{T\sqrt{k\overline{C}_V}}{\overline{E}}. \tag{5-24}$$

If this ratio is small enough, the probability of significant deviations from $\overline{E}$ in the value of E is negligible. Both C_V and E are extensive thermodynamic variables; for macroscopic systems they are proportional to the number of particles contained in the system. Thus the ratio in Eq. 5-24 varies inversely as $\sqrt{N}$. Furthermore; we can understand why the proportionality constant between Eq. 5-24 and $1/\sqrt{N}$ is not large from the case for ideal gases, which have C_V of order Nk and E of order NkT; thus Eq. 5-24 becomes simply of order $1/\sqrt{N}$. The average amount by which one would expect the measured energy of a system to differ from $\overline{E}$ is therefore something like $10^{-10}\overline{E}$ for macroscopic systems containing only 0.0001 mole, described by a canonical ensemble. It is inconceivable that such small fluctuations could have an effect big enough to show up on measurement.

*Andrews, Chap. 19.

The conclusion is that *one might as well use a canonical ensemble to represent a macroscopic system,* regardless of whether the system is isolated or in equilibrium with a heat bath. This is useful because in almost all applications the mathematics is much easier with the canonical distribution than with the microcanonical. Of course if the system under consideration is very small, Eq. 5-24 becomes appreciable. For example, suppose one were able to measure the energy contained in a volume of the size $10V/N$ available to 10 average molecules in an ideal gas. Then Eq. 5-24 takes a value near to $\frac{1}{3}$. For studying such small regions, having the whole spectrum of fluctuations is of much greater importance than simply knowing $\bar{E}$. This is characteristic of a physical world made up of particles. As the regions one considers become so small that they contain a relatively few particles, the simplification arising from averaging over the chaotic motion of vast numbers of particles is lost. This is discussed in some detail in Andrews, Chaps. 2 and 8. The average behavior *of a single particle* is worth very little as a prediction of what a given particle is doing; when we want to talk about single particles, we almost always look at the entire distribution function for the probabilities of interest. We return to the topic of fluctuations in Chap. 16.

It is an interesting question how the exponential damping of P_i with E_i, as in Fig. 5-2, can lead to such sharp peaking of probabilities around $\bar{E}$. There is nothing in the distribution

$$P_i = ae^{-\beta E_i} \tag{5-25}$$

or Fig. 5-2 to indicate such peaking. In fact, this shows that the states of lowest energy are most probable, and as E_i increases, the probabilities are damped. The answer is this: the probability Π_j that the system will be found *with a particular energy* E_j is the sum of the probabilities P_i of all states of energy E_j. The number of such states is the *degeneracy* W_j of the energy level E_j. Since all the W_j degenerate states have equal probabilities, $ae^{-\beta E_j}$, the probability Π_j that the system has energy E_j is the product of the degeneracy times the probability of a state:

$$\Pi_j = \sum_{\substack{\text{all states of} \\ \text{energy } E_j}} P_{\text{state}} = \sum_{\substack{W_j \text{ states of} \\ \text{energy } E_j}} ae^{-\beta E_j} = W_j ae^{-\beta E_j}, \tag{5-26}$$

$$\Pi_j = W_j P_i(E_j). \tag{5-27}$$

We now can see how it might be that the product of the exponentially damped function $P_i(E_j)$ (see Fig. 5-2) times the rapidly increasing function $W_i(E_i)$ (Fig. 4-2) might lead to an extremely sharply peaked dependence of

Π_j on E_j, as sketched in Fig. 5-3. For macroscopic systems, Fig. 5-3 is indistinguishable from Fig. 4-1.

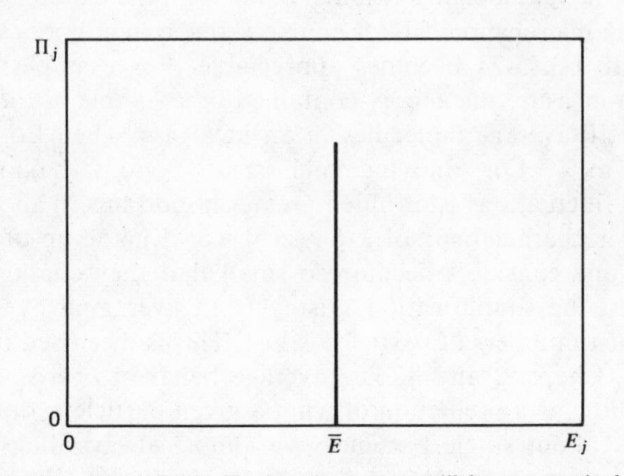

Fig. 5-3. Typical plot of probability Π_j that the system will have a particular energy as a function of energy E_j. This function is the product of the functions shown in Figs. 5-2 and 4-2. The total area under the curve is unity. For a macroscopic system of 10^{22} particles, the width of the peak is perhaps 10^{-11} times the distance of $\bar{E}$ from the origin.

Both the microcanonical and the canonical ensembles represent a system with a fixed number of particles N. Gibbs called these **petit ensembles,** in contrast to those in which the number of particles N is not fixed. The latter he called **grand ensembles,** and they can be considered to be made up of a large number of petit ensembles, weighted to give the desired dependence of the probability on the number of particles. Grand ensembles are considered in Chap. 16.

In concluding the discussion of the canonical distribution, it must be emphasized that the exponential form still may represent correlations between parts of the system represented by different ϵ's. These are correlations implicit in the meanings of the ϵ's, rather than explicit. For example, consider two particles. Of the group of ϵ's, one may be chosen to represent the potential energy of interaction between these particles. When the particles are close together, this value is large and affects markedly the probability of the configuration. When they are far apart, the value is small and can be neglected. Clearly the two particles are correlated, in that probabilities involving one of them depend on the position of the other through terms in the energy involving both their positions simultaneously. This type of correlation is discussed further in Chaps. 17 and 20.

In summary, the canonical distribution of probabilities of N-particle quantum states, as given in Eq. 5-9, is a reflection of the following necessary and sufficient conditions:

1. *The probability of an N-particle quantum state is a function of the energy of that state only.* Note: it must be meaningful to talk about separate quantum states for the system. If the system is interacting too strongly with its surroundings (e.g., the heat bath) for this condition to obtain, the theory breaks down.
2. *The system is only an infinitesimal part of a composite equilibrium situation consisting of system plus heat bath.* The heat bath serves only as an infinite source or sink of energy. If the system is sufficiently large, the heat bath is not even needed for the canonical ensemble to be a situable representation of an appropriate probability distribution.

PROBLEMS

5-1 Chapter 4 begins with a list of the information initially known about three different commonly encountered equilibrium conditions. Equations 4-5 and 5-9 give the probability distributions that represent the first two of these. Discuss for each condition how each item of initial information enters the probability distribution. Is there any information in the probability distributions that was not intially known?

5-2 Discuss the difference between

$$e^{-\Sigma_i \beta \epsilon_i} \quad \text{and} \quad \sum_i e^{-\beta \epsilon_i}$$

viewed simply as two abstract mathematical expressions.

5-3 The selection of a zero of energy is an arbitrary one. How does this choice affect the probability of a state in the canonical ensemble?

5-4 Sketch several plots similar to Fig. 5-2 but representing different values of β. Since $\beta = 1/kT$, make your sketches in the order of increasing temperatures.

6 THE ENTROPY IN STATISTICAL MECHANICS

FORMAL EXPRESSION FOR THE ENTROPY

A logical development of thermodynamics starts with a mechanical definition of *work* done on the system by the surroundings. Next it states the first law of thermodynamics, which permits work measurements in adiabatic processes to establish a function of state for the system which is called the *energy*, E. It continues by defining the *heat* absorbed by the system in a nonadiabatic process as the difference between its energy change and the work done on it in the process. It then states or obtains the zeroth law, which permits the consistent measurement of *temperatures*. Finally, it states the second law and establishes that there exists a universal function of only the empirical temperature that is an integrating factor for the reversible heat. That is, the heat $đq_{rev}$ in an infinitesimal reversible process is not the differential of a function of state, but this universal function of temperature *times* the reversible heat *is*. The reciprocal of the integrating factor is called the *absolute temperature* T, and $đq_{rev}/T$, known to be a perfect differential, is said to be the derivative dS of the state function *entropy* S. This development is summarized by the equations

$$đq = dE - đw, \tag{6-1}$$

$$dS = \frac{1}{T} đq_{rev}. \tag{6-2}$$

If Eqs. 6-1 and 6-2 are combined, one obtains

$$dE = T\,dS + đw_{rev}, \tag{6-3}$$

which expresses the combined first and second laws of thermodynamics.

*For example, see F. C. Andrews, *Thermodynamics—Principles and Applications* (Wiley-Interscience, New York, 1971), Chaps. 5–11.

We already have, in Eq. 5-14 or 5-15, a statistical mechanical expression for the energy of the system. To obtain an expression for the heat, we first need one for the work. Let a member of the ensemble be in the N-particle quantum state i with energy E_i. Suppose an infinitesimal amount of work is done on that member (e.g., by slightly changing the volume of the vessel containing it). This changes the energy eigenvalue for the ith state from E_i to $E_i + dE_i$. The work done on the member must of course be dE_i. In the entire ensemble, the members are in various states i with different values of E_i. The work dE_i on the members during the change in constraint (e.g., during the infinitesimal volume change) varies with the state i, but the average work over the whole ensemble is

$$\overline{đw} = \sum_i P_i dE_i, \tag{6-4}$$

from Eq. 2-28.

Now, having found the work, we can find the heat from Eq. 6-1:

$$\bar{E} = \sum_i P_i E_i, \tag{6-5}$$

$$d\bar{E} = \sum_i E_i dP_i + \sum_i P_i dE_i. \tag{6-6}$$

Since the last term in Eq. 6-6 is the average work $\overline{đw}$, the preceding term must be the average heat:

$$\overline{đq} = \sum_i E_i dP_i. \tag{6-7}$$

Now, to relate the heat to the entropy, it is most convenient to examine the derivative of the expression $\Sigma_i P_i \ln P_i$:

$$d\left(\sum_i P_i \ln P_i\right) = \sum_i \ln P_i dP_i + \sum_i dP_i \tag{6-8}$$

The last term in Eq. 6-8 is seen to be zero simply by differentiating the normalization condition:

$$\sum_i P_i = 1; \qquad \text{thus} \qquad \sum_i dP_i = 0. \tag{6-9}$$

Therefore, Eq. 6-8 becomes

$$d\left(\sum_i P_i \ln P_i\right) = \sum_i \ln P_i dP_i. \tag{6-10}$$

To go further, we must demand that the system be in equilibrium. Then we can determine $\ln P_i$ for a canonical ensemble:

$$P_i = \frac{e^{-\beta E_i}}{Q}; \qquad \ln P_i = -\beta E_i - \ln Q. \tag{6-11}$$

If this value is used in Eq. 6-10, we obtain

$$d\left(\sum_i P_i \ln P_i\right) = -\beta \sum_i E_i\, dP_i - \ln Q \sum_i dP_i. \tag{6-12}$$

Again, the last term is seen to vanish, by virtue of Eq. 6-9, and we are left with the result

$$d\left(\sum_i P_i \ln P_i\right) = -\beta \sum_i E_i\, dP_i. \qquad \text{(equilibrium ensemble)} \tag{6-13}$$

Comparison of this with Eq. 6-7 shows that for changes in which the system is always described by an equilibrium ensemble (noted by the subscript "rev"), we can write

$$\boxed{\overline{đq_{\text{rev}}} = -\frac{1}{\beta} d\left(\sum_i P_i \ln P_i\right).} \tag{6-14}$$

Using only the first law and the canonical ensemble, we have shown that the reversible heat indeed possesses an integrating factor β, which makes $\beta\, \overline{đq_{rev}}$ the differential of a function of state. This function is proportional to $\Sigma_i P_i \ln P_i$, a quantity that depends only on the number and kind of particles, the volume, and the temperature. This can be viewed, then, as a statistical mechanical derivation of this important thermodynamic result. To make the statistical mechanics consistent with the thermodynamics, the same definition of entropy must be made, namely, that of Eq. 6-2:

$$\overline{đq_{\text{rev}}} = T\, d\overline{S}. \tag{6-15}$$

Comparison of this with Eq. 6-14 shows that β must be inversely proportional to the absolute temperature. The proportionality constant is conventionally given the symbol $1/k$:

$$\boxed{\beta = \frac{1}{kT}.} \tag{6-16}$$

Choice of k essentially fixes the size and dimensions of the unit of absolute temperature. The scientific convention is to do this by reference to the properties of real gases in the limit of vanishing density.* This can be achieved theoretically by using statistical mechanics, a procedure followed in Chap. 10 (see Eq. 10-25). The result is the determination that k, called **Boltzmann's constant**, is the universal gas constant per particle, rather than per mole as used throughout thermodynamics; thus it is given by

$$\boxed{k=\frac{R}{N_0},} \tag{6-17}$$

where N_0 is Avogadro's number. From Eqs. 6-14 through 6-16, the statistical mechanical expression for the entropy is seen to be

$$dS=-k\,d\Big(\sum_i P_i \ln P_i\Big), \tag{6-18}$$

$$\boxed{S=-k\sum_i P_i \ln P_i,} \tag{6-19}$$

where for convenience the constant of integration was arbitrarily set at zero. This is appropriate, since that constant cannot depend on the quantum states available to the system or on the temperature. Any two states that could in principle be coupled by a reversible path would differ in entropy by an amount obtained by integrating Eq. 6-18 between the states; this amount would be identical to that calculated from Eq. 6-19 with the integration constant dropped. Since ΔS values are probably meaningless for two states where no reversible path coupling them can be imagined, dropping the integration constant is fully justified.†

Thus the entropy of a system, as viewed by statistical mechanics, is proportional to the ensemble average of the logarithm of the probability of an N-particle state:

$$\bar{S}=-k\,\overline{\ln P}\,. \tag{6-20}$$

*Andrews, Chaps. 9 and 14.

†For an interesting discussion of the integration constant, see E. Schrödinger, *Statistical Thermodynamics*, 2nd ed. (Cambridge University Press, Cambridge, 1952), pp. 15–17.

This average is a measure of how closely one can pinpoint the quantum state of the system. A high degree of predictive ability implies low entropy, and vice versa. The more quantum states are available to a system, the higher its entropy. This is a rather remarkable result. Equation 6-20 is an average unlike any others we have seen, since the average of ln P is not a *mechanical* quantity but an *informational* one, thus is a very human-centered quantity. We will spend considerable effort trying to gain an intuitive feeling for the entropy, because it is so important a thermodynamic concept, yet it proves so hard to understand.

Suppose the system is known with certainty to be in a particular N-particle quantum state j. Thus all the members in the ensemble must be in state j, and $P_j = 1$. The sum in Eq. 6-19 is thus reduced to the single term $-k(1 \ln 1)$, which is zero, since the logarithm of unity is zero.* Therefore, $\overline{S}$ is zero whenever the system is definitely known to be in a particular quantum state.

Of course we seldom know what N-particle quantum state the system is in. What we do know is either that the system is isolated with fixed energy or that it is in thermal equilibrium with a heat bath, thus has fixed temperature. In the first case, we know from Eq. 4-5 that P_i is $1/W$ for each of the W different degenerate N-particle states whose energy is E_{system}. In the second case, as we saw in the discussion accompanying Fig. 5-3, for macroscopic systems, the only states that contribute significantly to the ensemble are those whose energy is almost exactly $\overline{E}$. Although we do not know just how many such states these are, let us say that there are W of them. We do know that each of them has almost exactly the same probability, since their energies are all so near to one another:

$$P_i = ae^{-\beta E_i} \approx ae^{-\beta \overline{E}} = \frac{1}{W}, \tag{6-22}$$

where the last step followed by normalization. Thus whether the ensemble representing a macroscopic system is microcanonical or canonical, if the

*Since ln P blows up as P gets smaller, it is important to consider how $P \ln P$ behaves for very small P. This can be found from L'Hôpital's rule:

$$\lim_{P\to 0} P \ln P = \lim_{P\to 0} \frac{\ln P}{P^{-1}} = \frac{P^{-1}}{-P^{-2}} = -P. \tag{6-21}$$

Knowledge that contributions to the sum in Eq. 6-19 are of order P_i is useful later in deciding which states or energy levels make significant contributions to $\overline{S}$. Clearly, the limiting case where $P_i = 0$ need not concern us, since if there are no members of an ensemble in a state, that state does not enter into a summation of the form $\overline{g} = \Sigma_i P_i g_i$. Thus even if g_i were, say $1/i^2$, where the happy result of Eq. 6-21 did not hold, there would be no contribution to $\overline{g}$ from $i = 0$ because no member of the ensemble could have $i = 0$.

number of states in energy levels that contribute significantly to the ensemble is W, the probability of each of them is $1/W$. With this result, the summation of Eq. 6-19 is immediately performed:

$$\bar{S} = -k \sum_{i=1}^{W} \left(\frac{1}{W} \ln \frac{1}{W} \right), \tag{6-23}$$

$$= -kW \left(\frac{1}{W} \ln \frac{1}{W} \right), \tag{6-24}$$

$$= -k \ln \frac{1}{W}, \tag{6-25}$$

$$\boxed{\bar{S} = k \ln W.} \qquad \text{(macroscopic system)} \tag{6-26}$$

In Eq. 6-24, since each term in the sum is independent of the index i, the value of the sum is just the number of terms W times the value of one term. The entropy of a macroscopic system is k times the natural logarithm of the number of quantum states represented in the ensemble as significantly probable. This result, Eq. 6-26, is commonly called **Boltzmann's principle**; it is carved on the memorial where he is buried in Vienna.

Although Eqs. 6-19, 6-20, and 6-26 are commonly quoted and are useful in enhancing an intuitive feeling for entropy, they are not usually a practical means of finding the value of the entropy. For example, W is almost never known, except for special cases like those to be discussed. Statistical mechanics most commonly determines entropy from the partition function through Eq. 8-21, which follows directly from Eq. 6-19.

THE THIRD LAW OF THERMODYNAMICS

Consider the entropy of a system at a temperature of absolute zero. The probability $a e^{-\beta E_i}$ of a state is infinitely sensitive to the energy E_i, since β is infinitely large. Thus the system must be in the lowest possible energy level available to it. Given the degeneracy W of this level, the value of $\bar{S}$ at 0 K follows immediately from Eq. 6-26. For some models of crystals (called *perfect crystals*), quantum mechanical calculations show W to be unity, thus $\bar{S}$(0 K) is zero. A perfect crystal is one having no dislocations, no voids, no irregularities, no foreign atoms or molecules, no magnetic moment—nothing that would introduce the slightest element of choice among essentially equivalent quantum states, which would cause W to be larger than unity. There are, almost certainly, no perfect crystals, even if one could routinely reach 0 K.* The number of particles in a macroscopic system is so vast that to prepare a crystal with no flaws whatever is too unlikely ever to have happened. However, just how imperfect does a crystal have to be before Eq. 6-26 leads to a measurable entropy? It would represent extraordinary precision to be able to measure a molar entropy to within 10^{-5} eu. The maximum value of W that could show up in such a measurement can be calculated:

$$10^{-5}\,\frac{\text{cal}}{\text{deg}} = k \ln W = \frac{2.3\,R \log W}{N_0}, \tag{6-27}$$

$$\log W = \frac{(10^{-5}\ \text{cal})(6.02\times 10^{23})(\text{K-mole})}{\text{K(mole)}(2.3)(1.987\ \text{cal})}, \tag{6-28}$$

$$\log W \approx 10^{18}; \qquad W = 10^{10^{18}}. \tag{6-29}$$

In Eq. 6-27, we used Eq. 6-17. The result is rather astounding; *we do not need a perfect crystal to attain zero measurable entropy at* 0 K; *we only need a crystal whose degeneracy at* 0 K *is less than* $10^{10^{18}}$. That is a one followed by 10^{18} zeros; if printed in this size type, such a number would extend for about $\frac{1}{8}$ light year! This points up the extremely small size of k, or, correspondingly, the extremely large size of Avogadro's number.

For most crystals, the degeneracy of the lowest possible energy level, that achieved at 0 K, is well below anything that would lead to a measurable residual entropy. We thus have formulated, through statistical mechanics, a refinement of **Planck's statement of the third law of thermodynamics**, namely, that *as* $T \rightarrow 0$, *the entropy change approaches zero for any process for which a reversible path could be imagined, if the reactants and products are crystals whose degeneracies are less than about* $10^{10^{18}}$.

Planck's statement introduces a simplification into the thermochemical measurements needed to tabulate entropies of various compounds, a simplification that is not present in the case of enthalpies. Suppose one wanted to tabulate the enthalpy and the entropy of one mole of $CO_2(g)$ at 25 C. The state of zero enthalpy and entropy is conventionally taken to be

*That 0 K is unattainable experimentally is of no importance to this conclusion. Temperatures of 10^{-3} K are regularly achieved in laboratories, and 10^{-5} K has been reached. One feels confident about extrapolating crystal properties from these low values down to zero. Nothing unusual or unexpected is presumed to lie between 0.0000 and 0.0001 K.

the separated pure elements that make up the compound in question, at a temperature of 0 K. One could obtain $CO_2(g, 298\text{ K})$ from $C(s, 0\text{ K})$ and $O_2(s, 0\text{ K})$ by either of the following two processes:

1. $C(s,0\text{ K}) + O_2(s,0\text{ K}) \rightarrow CO_2(s,0\text{ K}) \rightarrow CO_2(g,298\text{ K})$,
2. $C(s,0\text{ K}) + O_2(s,0\text{ K}) \rightarrow C(s,298\text{ K}) + O_2(g,298\text{ K}) \rightarrow CO_2(g,298\text{ K})$,

and of course the same values of enthalpy and entropy must be found for both paths. For enthalpies, one regularly measures and tabulates ΔH for both steps in process 1 or process 2, whichever is most convenient. Indeed, ΔH for the chemical reaction step is less precisely measured than ΔH for the heating step. For entropies, however, Planck's statement assures us that the first step of process 1 has zero entropy change, since all pure crystals, be they elements or compounds, have zero entropy. Therefore, one needs to measure ΔS only for the heating step of process 1, which, happily, is the more precisely measured step anyway.

Interestingly enough, there *are* crystals having such huge degeneracies at 0 K that they do not follow Planck's statement. As an example, consider crystalline carbon monoxide. In the preparation of the crystal, while the molecules are freezing out, they have enough energy to make it immaterial in which direction they enter the lattice, CO or OC. Thus they enter randomly:

CO·CO·OC·OC
OC·CO·OC·OC ·
OC·OC·CO·OC

After the crystal is formed and the temperature greatly reduced, the energy differences between a perfect crystal,

CO·CO·CO·CO		CO·CO·CO·CO
OC·OC·OC·OC	or	CO·CO·CO·CO
CO·CO·CO·CO		CO·CO·CO·CO

and a random array may seem to be more important. But then it is too late for the CO molecules to turn around; there is too little energy available for them to squeeze past the barrier to rotation offered by their neighbors; they are frozen into their random orientation. The lowest energy level of the resulting crystal, which has randomly oriented molecules, may not be quite as low as that of a more regular crystal, but the difference is not likely to be great. But consider how this randomness on a molecular level affects the degeneracy of this level: specify the N-particle quantum state by giving the direction of orientation of the CO molecule at each of the N lattice points in the crystal. Each molecule introduces a factor of 2 into the total degeneracy; for example after one molecule has entered, the crystal has $W=2$, namely, $\uparrow$ and $\downarrow$. After 2 have entered, $W=4$, namely, $\uparrow\uparrow$, $\uparrow\downarrow$, $\downarrow\uparrow$, and $\downarrow\downarrow$; and so on: thus $W=2^N$. A number of this immensity leads to a measurable residual entropy:

$$\bar{S}(0\text{ K}) = k \ln 2^N = Nk \ln 2. \tag{6-30}$$

For a mole of CO, N is Avogadro's number, and this is 1.38 cal/mole-deg.

The residual entropy of CO has actually been measured as follows.* The entropy of gaseous CO at room temperature was calculated by the methods of Part 2 of this book, and the result agreed with the experimental value obtained by the two steps of the second process discussed previously for CO_2. However, the entropy found by integrating the experimentally determined $đq_{\text{rev}}/T$ from almost 0 K up to room temperature for CO (i.e., the second step of process 1 discussed for CO_2, above), proved to be less than the other by 1.0 cal/mole-deg. This suggests that the fraction 1.0/1.38 of the N particles actually entered the crystal as random units. Some of the N total may have been ordered in dimers when they entered the crystal; others may have had time to rearrange before the temperature got too low.

The foregoing comments can be generalized very easily. A particle may introduce a degeneracy in any of several ways. One is through its orientation, as with CO. Another is through its having several states for rotation (where the molecules still can rotate at almost 0 K), or several electronic states, all possessing the same low energy. If the crystal contains N_1 particles of type 1, each of which introduces degeneracy g_1, and N_2 particles of type 2 with degeneracy g_2, and so on, the residual entropy is

$$\bar{S}(0\text{ K}) = N_1 k \ln g_1 + N_2 k \ln g_2 + \cdots. \tag{6-31}$$

Examples of residual entropies other than that of CO are plentiful.† Nitrous oxide, N_2O, is a linear molecule and has residual entropy of $Nk \ln 2$. On the other hand, nitric oxide, NO, has experimental residual entropy of $\frac{1}{2}Nk \ln 2$. This indicates that just half as many particles as

*J. O. Clayton and W. F. Giauque, *J. Am. Chem. Soc.*, **54**, 2610 (1932).
†An extensive discussion is given by Fowler and Guggenheim, pp. 191–229.

expected are orienting randomly. One assumes that the NO molecules dimerize before crystallizing out as

$$\begin{matrix} \text{NO} \\ \text{ON} \end{matrix} \quad \text{or} \quad \begin{matrix} \text{ON} \\ \text{NO} \end{matrix} .$$

A crystal of CH_3D has residual entropy of $Nk \ln 4$, since there are four directions in which the deuterium may lie. This residual entropy is present with respect to the pure crystalline elements $C + \frac{3}{2}H_2 + \frac{1}{2}D_2$, where deuterium is treated as an element different from hydrogen. If one had no way to tell one isotope of an element from another, no entropy change could be perceived on mixing the isotopes, and no residual entropy would be found in a compound like CH_3D. The CH_3D crystal is of course the same in either case, but when isotopes are unobservable, the "pure elements" would be carbon plus a 3 : 1 mixture of hydrogen and deuterium.

Residual entropies are of course observed in solid solutions and in glasses because of the large degeneracies of these disordered substances.

PROBLEMS

6-1. Consider the residual entropy of carbon monoxide. Inability to predict the precise quantum state at 0 K led to a measurable residual entropy. However, one *possible* quantum state is the completely ordered state. If, unknown to the experimenter, the system had happened to reach that state, what would the experiment have shown? Does this mean that more heat would have to be added to raise that system to the same final condition described in the text? If so, where would the excess heat go? Considered in this light, why should measurements of residual entropies be reproducible?

6-2. Normal hydrogen (H_2) is a mixture of three parts of so-called *ortho*hydrogen ($g=3$) and one part of so-called *para*hydrogen ($g=1$). The residual entropy of hydrogen is therefore $\frac{3}{4}Nk \ln 3$. Using other references, find out what ortho and parahydrogen are, learn why they have these degeneracies, and determine the "pure elements" with respect to which this residual entropy is found.

6-3. Derive Eq. 8-21 for entropy from Eqs. 6-19, 6-5, and the definition of Q in Eq. 5-13.

6-4. In crystalline ice, each O atom is surrounded by four tetrahedrally located O atoms 2.76 Å away. Some place between each O–O pair lies an H atom. If the H's lay at 1.38 Å, what would be the residual entropy of ice? Ice has a residual entropy of 0.82 cal/deg-mole. This has been used as an argument that ice consists of discrete water molecules held in place by hydrogen bonds, with each O atom having two covalent H atoms at 0.99 Å and two hydrogen-bonding H atoms at 1.77 Å. If all these H_2O molecules were aligned in a definite pattern, what would be the residual entropy of ice? If these H_2O molecules aligned randomly, what residual entropy would result?

7 ENTROPY AND INFORMATION THEORY*

MEASUREMENT OF INFORMATION

"Information" is a word we use every day, intuitively and qualitatively. We now seek a definition precise enough to permit "information" to become the basis of a function of state; thus we will be able to talk about the "amount of information" as confidently as we talk about the "amount of energy."

First we must have a problem of well-defined dimensions—say, a system described by a set of properties whose range of possible values is known. Or, it might be a system with a known set of possible quantum states. Suppose we have an initial state of knowledge about those properties, although we do not know their exact values. Then we "receive a certain amount of information," which adds to our initial knowledge about the properties and leaves us in a state of final knowledge that is greater than the initial state. We seek a quantitative measure of the amount of information received. We will find that the amount is the difference between the values of two state functions derived from the initial and final states of knowledge. This is fortunate, since we will deduce the same amount of "information received," whether it comes in several installments, one after the other, or all at once. This is suggested by Fig. 7-1.

Of the set of permitted events (or values of a property, or states of a system), suppose initially we know only that G of them are equally possible and the rest are impossible. The initial field of uncertainty consists of G possible events, each with equal probability $P^0 = 1/G$. Then let us define one **binary unit of information**, the so-called **bit**, to be *the amount of information that permits us to make a choice between equally probable events.* One bit of information is received whenever a question is answered that

*This chapter is included because of the intuitive light it sheds on the meaning of entropy and because information theory is an interesting discipline in its own right. It may be omitted from a reading of the book with almost no loss to comprehension of the rest of the chapters.

allows us to rule out all members of the ensemble that contribute exactly $\frac{1}{2}$ to the total probability distribution being considered. Thus after one bit is received, events that previously had a total probability of $\frac{1}{2}$ are known to be impossible; this simultaneously doubles the probabilities in the new ensemble of all the events still possible. In our example, after one bit is received, only $\frac{1}{2}G$ events are known to be possible (equally so); after 2 bits are received, $\frac{1}{4}G$ events are known to be possible (equally so); after I bits are received, $(\frac{1}{2})^I G$ events are known to be possible (equally so). Each bit received narrows the field of uncertainty by exactly half. Thus after asking I binary questions, the field of possible events is reduced from G to H, each with probability $P' = 1/H$, the probability of the other $G - H$ events being reduced to zero.

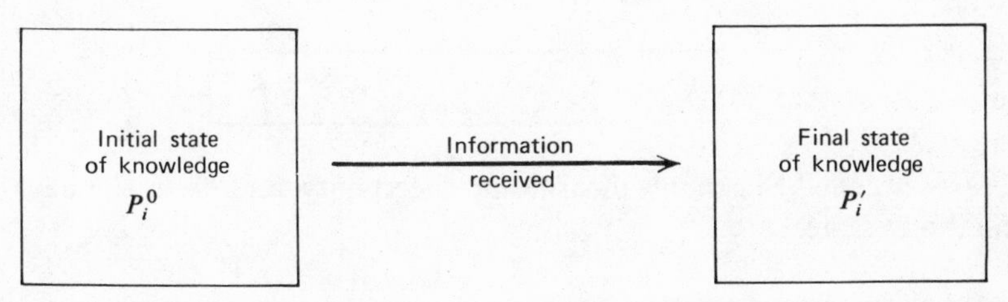

Fig. 7-1. Two states of knowledge about a well-defined problem. Each state is encoded in an ensemble representing what is known; this generates the probability distributions P_i^0 and P_i'.

Mathematically, this becomes

$$\left(\frac{1}{2}\right)^I G = H \qquad \text{or} \qquad \frac{G}{H} = 2^I. \tag{7-1}$$

We may take the logarithm of both sides:

$$\ln\frac{G}{H} = I \ln 2 \tag{7-2}$$

or

$$I = -\frac{\text{bit}}{\ln 2}\ln H + \frac{\text{bit}}{\ln 2}\ln G. \tag{7-3}$$

Now we may substitute the values $P^0 = 1/G$ and $P' = 1/H$:

$$I = \frac{\text{bit}}{\ln 2}\ln P' - \frac{\text{bit}}{\ln 2}\ln P^0. \tag{7-4}$$

Finally, we may exploit the fact that *each* of the nonzero probabilities has the same value, either $P_i' = 1/H$ or $P_i^0 = 1/G$; thus we write

$$\ln P_i' = \ln P_i' \sum_i P_i' = \sum_i P_i' \ln P_i'. \tag{7-5}$$

This is analogous to a procedure obtaining the form of Eq. 6-19 for equally probable events by proceeding in reverse order, starting with Eq. 6-26. Use of Eq. 7-5 in Eq. 7-4 yields

$$I = \frac{\text{bit}}{\ln 2} \sum_i P_i' \ln P_i' - \frac{\text{bit}}{\ln 2} \sum_i P_i^0 \ln P_i^0. \tag{7-6}$$

The result, Eq. 7-6, shows the information received, as measured in bits, to be the difference between two functions of the state of knowledge of the system:

$$\boxed{I = \Delta \mathfrak{N} = -\Delta \mathfrak{S} = \mathfrak{N}' - \mathfrak{N}^0 = \mathfrak{S}^0 - \mathfrak{S}',} \tag{7-7}$$

where $\mathfrak{S}$ is called **Shannon's measure of uncertainty**, and $\mathfrak{N}$ is the negative of Shannon's measure:

$$\boxed{\mathfrak{S} \equiv -\frac{\text{bit}}{\ln 2} \sum_i P_i \ln P_i; \qquad \mathfrak{N} \equiv \frac{\text{bit}}{\ln 2} \sum_i P_i \ln P_i.} \tag{7-8}$$

We have proved this only for ensembles describing equally probable events, but the same result can be shown to hold for arbitrary ensembles. The value of Shannon's measure in bits is the number of binary decisions needed to remove all uncertainty from one's knowledge, thus to *determine* which event will occur (i.e., to reduce all P_i to 0 except for one that has $P_j = 1$).

For example, consider the simple choice between equally probable events that reduces the initial probability distribution

$$P_1^0 = \tfrac{1}{2}; \qquad P_2^0 = \tfrac{1}{2} \tag{7-9}$$

to a certainty:

$$P_1' = 1; \qquad P_2' = 0. \tag{7-10}$$

The value of I is obtained from Eq. 7-6:

$$I = \frac{\text{bit}}{\ln 2}(1 \ln 1 + 0 \ln 0) - \frac{\text{bit}}{\ln 2}\left(\frac{1}{2}\ln\frac{1}{2} + \frac{1}{2}\ln\frac{1}{2}\right), \tag{7-11}$$

$$I = 1\,\text{bit}. \tag{7-12}$$

This is completely in accordance with our definition of I. *The value of $\mathcal{S}$ is the number of binary questions that must be answered to reduce an uncertain state of knowledge to a certainty.* Thus one can meaningfully compare the amount of information or uncertainty in two different ensembles or probability distributions; the one with greater $\mathcal{S}$ has less information, since it would take more binary questions to reduce it to a certainty.

ENSEMBLES WITH MINIMUM INFORMATION

Now that we can measure the information contained in an ensemble and we know it to be meaningful and consistent to talk about this amount of information, we are in a position to ask the following question. Given a set of states i, what assignment of probabilities P_i reflects maximum ignorance or minimum information about the likelihood of those states? This problem reduces to the mathematical task of maximizing Shannon's measure, or finding the extremum of

$$\mathcal{G} = \sum_{i=1}^{G} P_i \ln P_i \tag{7-13}$$

subject to the normalization condition

$$\sum_{i=1}^{G} P_i = 1. \tag{7-14}$$

The extremum of $\mathcal{G}$ is the value of $\mathcal{G}$ when its derivatives with respect to all the independent variables P_i vanish. We thus want to differentiate $\mathcal{G}$ with respect to the independent P's, but because of Eq. 7-14, not all the P's are independent. One of them must be viewed as being fixed by Eq. 7-14. Let us choose P_G to be this particular one:

$$P_G = 1 - \sum_{i=1}^{G-1} P_i. \tag{7-15}$$

Then all the other P's can be varied independently. Since we want to maximize

$$\mathcal{S} = \sum_{i=1}^{G-1} P_i \ln P_i + P_G \ln P_G, \tag{7-16}$$

we differentiate $\mathcal{S}$ with respect to one of the independent P's, say, P_j, and set the derivative equal to zero:

$$\frac{\partial \mathcal{S}}{\partial P_j} = 0 = \frac{\partial}{\partial P_j}(P_j \ln P_j + P_G \ln P_G), \tag{7-17}$$

where it was clear that the only term in the sum in Eq. 7-16 that contains P_j is $P_j \ln P_j$. Differentiating the two products yields

$$0 = 1 + \ln P_j + 1\frac{\partial P_G}{\partial P_j} + \ln P_G \frac{\partial P_G}{\partial P_j}. \tag{7-18}$$

From Eq. 7-15 we see that

$$\frac{\partial P_G}{\partial P_j} = -1, \tag{7-19}$$

and Eq. 7-18 becomes simply

$$\ln P_j - \ln P_G = 0 = \ln \frac{P_j}{P_G}. \tag{7-20}$$

The only quantity whose logarithm is zero is unity; thus we have

$$P_j = P_G \qquad \text{for all } j. \tag{7-21}$$

Thus *all the P's must have the same value* at the maximum of $\mathcal{S}$: $P_i = 1/G$ for all i maximizes the uncertainty. This surely agrees with our intuitive understanding that if we were to weight P_i differently from P_j in the ensemble, we would require a *reason* to do so (i.e., some *information* that suggests doing so). This has already been discussed in connection with Eq. 4-1 for the microcanonical ensemble, where all quantum states consistent with the information available were weighted equally in the ensemble.

ENTROPY AND INFORMATION

The fascinating result of the foregoing findings is that the entropy as expressed by statistical mechanics is proportional to Shannon's $\mathbb{S}$, thus is a measure of uncertainty regarding which quantum state the system is in. Comparison of Eqs. 6-19 and 7-8

$$S = -k\sum_i \cdots \qquad \text{and} \qquad \mathbb{S} = -\frac{\text{bit}}{\ln 2}\sum_i \cdots \tag{7-22}$$

shows that the entropy is

$$S = \frac{k \ln 2}{\text{bit}}\,\mathbb{S}, \tag{7-23}$$

where $\mathbb{S}$ *is the number of binary decisions needed to determine which quantum state the system is in.* For S given in entropy units (eu$\equiv$cal/deg),

$$S(\text{in eu}) = \frac{R}{N_0}\frac{\ln 2}{\text{bit}}\,\mathbb{S} = \frac{(1.987\,\text{eu})(\text{mole})(2.3)\log 2\ \mathbb{S}}{(\text{mole})(6.02\times 10^{23})\text{bit}}, \tag{7-24}$$

$$S(\text{in eu}) = 2.29\times 10^{-24}\left(\begin{array}{l}\text{number of binary decisions needed to deter-}\\ \text{mine which quantum state system is in}\end{array}\right), \tag{7-25}$$

$$\left(\begin{array}{l}\text{Number of binary decisions needed to deter-}\\ \text{mine which quantum state system is in}\end{array}\right) = 4.37\times 10^{23} S(\text{in eu}).$$

Any intuitive aid in understanding the mysterious but ubiquitous function, entropy, is grounds for rejoicing. For example, consider the residual entropies of crystals at 0 K discussed in Chap. 6. The entropy will be unmeasurably small (less than 10^{-5} eu) in any system for which the number of binary decisions needed to determine the quantum state of the system is less than 4.4×10^{18}. An example is the set of quantum states available to a carbon monoxide crystal at 0 K, when all molecules plus the crystal as a whole are in their ground vibrational states. Suppose, however, that at each lattice site, the orientation of each molecule CO or OC is absolutely random. To specify the state of the crystal as a whole, one would have to go from lattice site to lattice site, stating whether the carbon monoxide there was CO or OC. The randomness of the orientation means that a priori the likelihood is 50–50, thus specification of the orientation at each lattice site requires one bit of information. Since there are N separate sites in the crystal, $\mathbb{S}$ is N bits, and Eq. 7-23 is in complete agreement with Eq. 6-30 for the residual entropy. Since actual crystals of CO show residual entropies of 1.0/1.38 of this amount, it appears that the occupation of lattice sites is not random; knowledge of the orientations in neighboring sites changes the a priori probability from $P_{CO}=P_{OC}=\frac{1}{2}$ to something else, for which of course less than one bit of information (1.0/1.38 bit, on average) is required to reduce to certainty.

Another example clarified by the information theory picture of entropy is the dependence of S on V for ideal gases. Suppose N molecules are known initially to be in the volume V_1 and finally to be in the volume V_2, where $V_2 < V_1$. The field of uncertainty in the position of each molecule is reduced by a factor of V_2/V_1, and the information gained about each molecule is therefore found from Eq. 7-3 to be

$$I=\frac{\text{bit}}{\ln 2}\ln\frac{G}{H}=\frac{\text{bit}}{\ln 2}\ln\frac{V_1}{V_2}\,. \qquad (7\text{-}26)$$

The total gain in information is N times this quantity, and the entropy change, using Eqs. 7-7 and 7-23, is

$$\Delta S=\frac{k\ln 2}{\text{bit}}\Delta\mathbb{S}=-\frac{k\ln 2}{\text{bit}}I=-Nk\ln\frac{V_1}{V_2}\,. \qquad (7\text{-}27)$$

This agrees precisely with the dependence of S on V derived thermodynamically for ideal gases at constant T.

A final example of entropy changes clarified by information theory is the so-called entropy of mixing of ideal gases at constant total pressure and temperature. The opposite of the mixing process is the separation or sorting, depicted in Fig. 7-2. The field of uncertainty in the position of each molecule is reduced by a factor of $\frac{1}{2}$; thus the information gained is one bit for each of the N molecules. These N bits represent, from Eq. 7-27, an entropy change of $-Nk\ln 2$; thus the mixing, which is the opposite of the process considered here, must increase S by $Nk\ln 2$. This also agrees with the thermodynamic result.

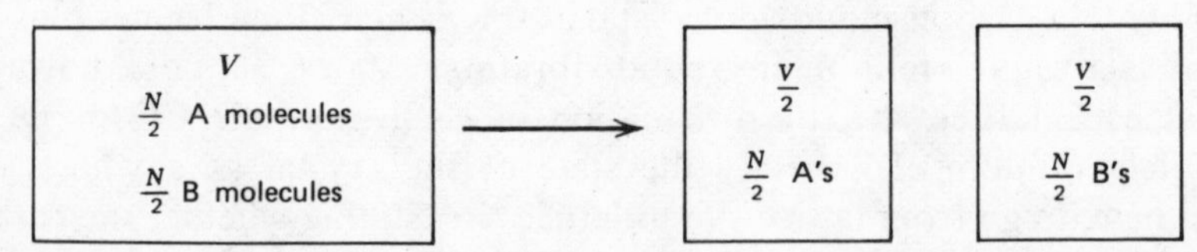

Fig. 7-2. Process that separates a mixture of type A and B molecules in an ideal gas at constant temperature and total pressure.

This example is a classical illustration of the equivalence between information and negentropy. (**Negentropy** is the negative of the entropy.) A little creature (Maxwell's demon) capable of distinguishing between A and B molecules as they came leaking out of the original volume could shunt the A's into one vessel and the B's into another, achieving the result entirely adiabatically at constant total pressure, total volume, and temperature. Such sorting, viewed from the surroundings, has the appearance of being reversible, yet adiabatic reversible processes should not decrease the entropy. The conclusion, borne out by careful analysis, is that in the process of identifying the A's and B's and in manning the trapdoors that shunt them into the proper vessels, the demon must expend an entropy of at least $Nk\ln 2$. The demon itself and the aids it uses to accomplish its work must end up having run down sufficiently to pay for the winding up (or increase in order) represented by the gas separation. The second law is not violated by the presence within systems of agents that obtain and act on information. Isolated systems still run down and reach equilibrium. However, the information gathering and processing agents (such as living organisms) and all their aids must be considered to be part of the system. Otherwise, if they are taken to be part of the surroundings, their injections of information are the same as additions of negentropy from the outside, one bit of information being equivalent to $k\ln 2$ of negentropy.

PROBLEMS

7-1. How much information is obtained from each roll of an unbiased die? From each roll of a pair of such dice, assuming the dice are distinguishable? How is the latter question answered in the more usual event that the dice are indistinguishable?

7-2. Suppose one is told that a die, previously thought to be honest, was loaded so that 1 appeared twice as often as any of the other five numbers, which were otherwise equally probable. How much information was contained in the message? Note how information measured in bits ignores the importance of the message, which might be almost nothing in a friendly Monopoly game but might mean life or death in certain crap games.

7-3. We toss three coins labeled a, b, and c, and report the following to four friends about the results of the experiment: (1) the coins are not all three heads, (2) coin a is a tail, (3) there are two tails and one head, (4) coin b is a head and a and c are tails. How many bits of information are conveyed to each friend? Suppose we made the same four reports, in the same order, to a single fifth friend. How many bits would be conveyed in each report? Is the total information received by the fifth friend the same as what the fourth friend previously received? Why?

7-4. Suppose the center of mass of a particle, initially known only to be within the region $x=0$ to $x=L$, is located as the result of an experiment and found to be within a region of length Δx. How much information is obtained by the experiment? How much negentropy must be expended in measuring this position? If the center of mass were located exactly, how much information would be obtained? What would be the cost in negentropy? Might this result be related to the need to introduce quantum mechanics with its Heisenberg uncertainty principle into physics to replace Newtonian classical mechanics?

7-5. Most of classical mathematics is based on the properties of points, lines, and surfaces. Does Prob. 7-4 suggest any limitations on the applicability of classical mathematics to relating the measured properties of systems?

7-6. The message carried in the genetic DNA of living organisms is written in a language containing four different nucleotide pair letters. Specification of each single letter thus reduces the field of uncertainty by $\frac{1}{4}$. A typical sperm cell contains a strand of DNA in which are preserved about 10^{10} bits of information. How many nucleotide pair letters long is the DNA message in such a cell?

7-7. The proteins of living organisms may be viewed as messages, written by the DNA, in a language containing 20 different amino acid letters. How much information is contained in the specification of each amino acid residue? What is the minimum number of nucleotide pairs in DNA that could code for specifying each amino acid residue in a protein? Nature chose to require three such pairs; is that number consistent with your findings?

7-8. As information is defined, it measures the unexpectedness of a message in a situation in which all possible messages are in principle capable of enumeration and an a priori probability is assigned to each. What limitations are apparent on the utility of information defined in this way?

7-9. Prove that Shannon's measure of uncertainty based on the joint probability P_{ij} of two events equals the sum of the uncertainties of each event separately (based on P_i and P_j) whenever events i and j are uncorrelated.

7-10. Noting the definition of $\mathbb{S}$ following Eq. 7-12 and the heuristic expression for S, Eq. 6-26, derive Eq. 7-24 for S directly from a starting point of the form of Eq. 7-1. This shows the simple equivalence of Eqs. 6-26 and 7-24. It also suggests that much of the enthusiasm over the prospective use of information theory in understanding microscopic events is likely to be misplaced, since Eq. 6-26 has been with us since Boltzmann first suggested it in 1877 and Planck first wrote it in 1906.

ADDITIONAL READING

L. Brillouin, *Science and Information Theory*, Academic Press, New York, 2nd ed., 1962.

H. A. Johnson, "Information Theory in Biology after 18 Years," *Science*, **168**, 1545 (1970).

M. Tribus and E. C. McIrvine, "Energy and Information," *Scient. Am.*, **224**, 179 (1971).

8 GENERAL FORMULATION OF STATISTICAL THERMODYNAMICS

Thermodynamics shows that all the various equilibrium properties of a system can be found immediately from knowledge of a so-called thermodynamic potential as a function of the proper set of variables for that potential.* The most commonly used potentials, with the correct independent variables for each, are the energy $E(S,V,N)$, the enthalpy $H(S,p,N)$, the Helmholtz free energy $A(T,V,N)$, and the Gibbs free energy $G(T,p,N)$. Here, N represents the number of particles in the system.

The development in Chaps. 5 and 6, based on the canonical ensemble, is concerned with a system about which the following are known: the number N and kind of particles, the volume V and shape of the vessel containing the system, and the temperature T (or the parameter β, proved in Chap. 10 to be $1/kT$). All these data except T are used in setting up and solving the Schrödinger equation, a process that yields the set of allowed quantum states for the system, with the appropriate energy eigenvalues. The temperature describes the heat bath with which the system is in equilibrium. It enters in the weighting of the relative probabilities of the resulting quantum states in the canonical ensemble. Thus the natural independent variables with the canonical distribution are T, V, and N. We seek therefore an expression for the ensemble average of A.† This is found immediately from the definition of A,

$$A \equiv E - TS, \tag{8-1}$$

from Eqs. 2-28 and 6-19 for E and S,

$$E = \sum_i E_i P_i; \qquad S = -k \sum_i P_i \ln P_i, \tag{8-2}$$

*Andrews, Chap. 13 and Appendix B.

†Throughout the rest of the book we drop the overbar notation for the ensemble average of thermodynamic quantities whenever there is no doubt that such averages are implied.

which yields

$$A=\sum_i (E_i+kT\ln P_i)P_i, \tag{8-3}$$

and from Eq. 6-11 for ln P_i, which yields

$$A=\sum_i (E_i-kT\beta E_i-kT\ln Q)P_i. \tag{8-4}$$

Since β is $1/kT$, the first two terms between the parentheses cancel. The $-kT\ln Q$ comes outside the summation, the summation is simply unity because of normalization, and we are left with

$$\boxed{A=-kT\ln Q.} \tag{8-5}$$

This most important equation sums up all statistical thermodynamics in terms of the partition function

$$\boxed{Q\equiv\sum_i e^{-\beta E_i}.} \tag{8-6}$$

This sum over all N-particle quantum states of the exponential in the energy of the state (some books use the symbol Z rather than Q) has a central role in statistical thermodynamics, since once it is known as a function of T, V, and N, all thermodynamic quantities can be calculated from it directly. Planck called it the *Zustandssumme*, and for a while (Tolman) there was hope that the natural translation, *sum-over-states*, would become its English name. However, the name **partition function**, used by Darwin and Fowler for an analogous quantity, has become generally accepted for Q.

The actual derivation of the various thermodynamic quantities as functions of T, V, and N from

$$A=A(T,V,N_1,N_2,\ldots) \tag{8-7}$$

is performed as follows: if a system consisting of N_1 particles of type 1, N_2 particles of type 2,..., is caused to undergo infinitesimal changes in temperature, volume, and numbers of particles, the change in Helmholtz free energy is given by

$$dA=\left(\frac{\partial A}{\partial T}\right)_{V,N_i}dT+\left(\frac{\partial A}{\partial V}\right)_{T,N_i}dV+\sum_i\left(\frac{\partial A}{\partial N_i}\right)_{T,V,N_{j\neq i}}dN_i. \tag{8-8}$$

The summation is over all types of constituent particles. The meaning of the symbol $\partial A/\partial N_i$ is the change in A caused by increasing the number of particles of type i from N_i to N_i+1. The meaning of the symbol dN_i is the actual change in the number of particles of type i. The $N_{j\neq i}$ means all particle numbers N_j are held constant except for the ith, which is explicitly changing.

This purely mathematical equation can be compared with its thermodynamic counterpart, one of the so-called *Gibbs equations:*

$$dA = -S\,dT - P\,dV + \sum_i \mu_i\,dN_i. \tag{8-9}$$

The symbol μ_i is the *molecular chemical potential* or *partial molecular Gibbs free energy* of the ith constituent. It differs from the molar chemical potential or partial molar free energy customarily used by chemists in being smaller by a factor of Avogadro's number. It is consistently used in this way throughout this book, because the emphasis in statistical mechanics is on the particulate nature of matter.

Comparing Eq. 8-8 with Eq. 8-9 permits the following identifications:

$$S = -\left(\frac{\partial A}{\partial T}\right)_{V,N_i}; \qquad P = -\left(\frac{\partial A}{\partial V}\right)_{T,N_i}; \qquad \mu_i = \left(\frac{\partial A}{\partial N_i}\right)_{T,V,N_{j\neq i}}. \tag{8-10}$$

Coupled with the simple form of A, Eq. 8-5, Eqs. 8-10 give simple ways to remember S, P, and μ_i. In addition, E is given by

$$E = A + TS; \tag{8-11}$$

H and G are given by their definitions:

$$H \equiv E + PV; \tag{8-12}$$

$$G \equiv A + PV; \tag{8-13}$$

and the heat capacities are given by differentiation of E and H:

$$C_V = \left(\frac{\partial E}{\partial T}\right)_{V,N_i}; \qquad C_P = \left(\frac{\partial H}{\partial T}\right)_{P,N_i}. \tag{8-14}$$

At this point we summarize the ensemble averages of the various thermodynamic variables.*

*Note, for any function of temperature $g(T)$, we can put

$$\frac{\partial g}{\partial T} = \frac{\partial g}{\partial \beta}\frac{d\beta}{dT} = \frac{1}{k}\frac{\partial g}{\partial \beta}\frac{d(1/T)}{dT} = -\frac{1}{kT^2}\frac{\partial g}{\partial \beta}. \tag{8-15}$$

$$E = kT^2\left(\frac{\partial \ln Q}{\partial T}\right)_{V,N_i}; \tag{8-16}$$

$$H = kT^2\left(\frac{\partial \ln Q}{\partial T}\right)_{V,N_i} + kTV\left(\frac{\partial \ln Q}{\partial V}\right)_{T,N_i}; \tag{8-17}$$

$$A = -kT\ln Q; \tag{8-18}$$

$$G = -kT\ln Q + kTV\left(\frac{\partial \ln Q}{\partial V}\right)_{T,N_i}; \tag{8-19}$$

$$\mu_i = -kT\left(\frac{\partial \ln Q}{\partial N_i}\right)_{T,V,N_{j\neq i}} = -kT\ln\frac{Q_{N_i+1}}{Q_{N_i}}; \tag{8-20}$$

$$S = k\left[\frac{\partial (T\ln Q)}{\partial T}\right]_{V,N_i} = k\ln Q + \frac{E}{T}; \tag{8-21}$$

$$C_V = \left(\frac{\partial E}{\partial T}\right)_{V,N_i}; \tag{8-22}$$

$$C_P = \left(\frac{\partial H}{\partial T}\right)_{P,N_i}; \tag{8-23}$$

$$P = kT\left(\frac{\partial \ln Q}{\partial V}\right)_{T,N_i}. \tag{8-24}$$

The second form of Eq. 8-20 follows thus:

$$\frac{\partial \ln Q(N_i)}{\partial N_i} = \frac{\ln Q(N_i+1) - \ln Q(N_i)}{1} = \ln\frac{Q(N_i+1)}{Q(N_i)}. \tag{8-25}$$

These equations show why the partition function is so widely used to calculate thermodynamic variables from properties of the molecules. Whenever one has a system described by a canonical ensemble, if $e^{-\beta E_i}$ can be summed conveniently over all states to yield Q, the values of all thermodynamic variables follow immediately. Needless to say, this has not solved the problem of equilibrium statistical mechanics. It has, however, established a framework, throwing the entire problem into the calculation of the one quantity Q as a function of the variables on which it depends. Much of the rest of this book is devoted to studying this problem for various kinds of system. Finding Q for systems other than ideal gases is by no means easy; in most cases it represents an unsolved research problem.

To provide enhanced intuitive appreciation for Q, it is worth relating Q to the number of states W that contribute significantly to the ensemble:

$$A = E - TS = -kT\ln Q. \tag{8-26}$$

Using Eq. 6-26 for S yields

$$E - kT\ln W = -kT\ln Q \tag{8-27}$$

or

$$Q = We^{-E/kT}. \qquad \text{(macroscopic system)} \tag{8-28}$$

Thus Q is a measure of the number of quantum states in which the system might be found, damped by the exponential in the energy of the system. This result follows immediately from consideration of Eq. 8-6, since almost all the states contributing significantly to the sum over states have very nearly the same energy E.

Again, however, we note that equations like 8-28 involving W have little practical application, since W is so rarely known. Instead, one must calculate Q from Eq. 8-6. Another approach taken to finding Q is to determine the **density of quantum states** $\rho(V,N,E)$ defined as follows: $\rho(V,N,E)\,dE$ is the number of quantum states for the N-particle system whose energy lies between E and $E+dE$. From the density of states, Q follows immediately:

$$Q(T,V,N) = \int_0^{\infty} \rho(V,N,E)e^{-E/kT}\,dE, \tag{8-29}$$

which is completely equivalent to Eq. 8-6.*

Finally, it is interesting to note that the probability that the system is in its ith quantum state can be written in the following, often encountered, form:

$$P_i = Q^{-1}e^{-\beta E_i} = e^{\beta(A - E_i)}. \tag{8-30}$$

*For people familiar with the concept, this shows Q to be the Laplace transform of the density of states.

PROBLEMS

8-1. Consider a hypothetical system having only three accessible quantum states, a ground state of energy zero, and two degenerate excited states of energy ϵ. This system is in equilibrium with a heat bath of temperature T. Calculate Q, A, E, S, and C_V. What values would the probabilities of each of the three states tend toward as T became very small? What values would these probabilities tend toward as T became very large? When one talks about high or low temperatures, he means high or low with respect to what quantity?

What would you expect a plot of $\bar{E}$ versus T to look like? What would you expect to plot of $\overline{C_V}$ versus T to look like? If the zero of energy is arbitrarily chosen to be the level of the two excited states, how will that choice affect the values of Q, A, E, S, and C_V?

8-2. Assume that some quantum state for the system has zero energy.

PART 2

QUANTUM STATISTICAL THERMODYNAMICS OF IDEAL GASES

9 FERMI–DIRAC, BOSE–EINSTEIN, AND BOLTZMANN STATISTICS

PARTICLE INDISTINGUISHABILITY

The first problem in determining the partition function,

$$Q = \sum_{\substack{\text{all } N\text{-particle} \\ \text{quantum states}}} e^{-\beta E_{\text{state}}}, \tag{9-1}$$

is determining the complete set of N-particle quantum states and corresponding energies, that is, solving Eq. 3-1 for the N-particle system, subject to the appropriate boundary conditions. We shall call an **ideal gas** one in which there is no mechanical interaction between the particles. Each particle is described quantum mechanically as if it were alone in the box. Thus each particle is in one of the available single-particle quantum states, a solution of the much simpler Schrödinger equation,

$$\mathfrak{h}\psi_i(\mathbf{r}) = \epsilon_i \psi_i(\mathbf{r}). \tag{9-2}$$

The difference between Eqs. 9-2 and 3-1 is significant. The wave function ψ_i in Eq. 9-2 describes a single particle in the ith particle quantum state, and ψ_i is a function of the coordinates of one particle. The Hamiltonian operator $\mathfrak{h}$ is obtained by the usual prescription from the classical expression for the total energy of that single particle. The energy eigenvalue ϵ_i is the energy of the single particle when it is in particle quantum state i.

For ideal gases, the solution of Eq. 3-1, the Schrödinger equation for the system, which is needed for the evaluation of Eq. 9-1, is usually reducible to the problem of solving Eq. 9-2 for a single particle. The only question remaining is how to specify an N-particle state in terms of the single-particle states of the N constituent particles. To this question, quantum mechanics responds as follows: *a quantum state for the N particles of an ideal gas is determined by giving the numbers of particles in each available particle quantum state.* Only the numbers of particles in each particle state are needed; there is no meaning attached to which of a group of identical particles is in which state. This is because the identical particles that make up a gas or liquid are completely indistinguishable. Hence the quantum state of the N-particle system is not changed if two identical particles are simply interchanged. Thus the set of **occupation numbers,** that is, the numbers of particles occupying the various particle states, completely specifies an N-particle quantum state.

It is a general conclusion of quantum mechanics that *all* particles are one or the other of two kinds:* if the interchange of two identical particles leads to an N-particle wave function that is the negative of its value before interchange, the particles are called **fermions** and are said to obey **Fermi-Dirac statistics.** No two fermions can be in the same particle state; thus the only possible value for the occupation numbers of fermions is 0 or 1. Identical particles for which the interchange of two of them leads to an N-particle wave function that is the same as its value before interchange are called **bosons** and are said to obey **Bose-Einstein statistics.** Since any number of bosons can be in the same particle state, occupation numbers for bosons can take on any integral value between 0 and N. Of course, the interchange of two bosons, like the interchange of two fermions, does not lead to a new N-particle state.

Experiments have determined which particles are fermions and which are bosons.† All particles whose spins are odd multiples of $\frac{1}{2}$ prove to be fermions. Examples of fundamental particles with such spins are electrons, neutrons, and protons. All particles whose spins are even multiples of $\frac{1}{2}$ prove to be bosons. Photons are fundamental particles with integral spin. To determine which statistics will be obeyed by an arbitrary molecule, ion,

*Tolman, Sec. 76; Schiff, Sec. 32; Heitler, Chap. V; d'Abro, pp. 931–943. The reader familiar with quantum mechanics will appreciate the following argument. The permutation operator P_{12} interchanges the dependence of the N-particle wave function Ψ on the positions of particles 1 and 2: $P_{12}\Psi(r_1,r_2,\cdots)=\Psi(r_2,r_1,\cdots)$. If P_{12} is applied twice in succession onto Ψ, the original value of Ψ must of course be recovered, since two permutations leaves the system unchanged. Thus the square of the eigenvalue λ_{12} of P_{12}, defined by $P_{12}\Psi(r_1,r_2,\cdots)=\lambda_{12}\Psi(r_1,r_2,\cdots)$ must be unity. Since $\lambda_{12}^2=1$, λ_{12} must be either $+1$ or -1. If one pair of identical particles has $\lambda_{12}=+1$, then all pairs of similar particles must have $\lambda_{ij}=+1$; similar consistency holds for identical particles for which $\lambda_{ij}=-1$. Proof of that follows thus: since both the product operators $P_{23}P_{12}$ and $P_{13}P_{23}$ take $\Psi(r_1,r_2,r_3,\cdots)$ into $\Psi(r_2,r_3,r_1,\cdots)$, they both must have the same eigenvalue. these eigenvalues are $\lambda_{23}\lambda_{12}$ for $P_{23}P_{12}$ and $\lambda_{13}\lambda_{23}$ for $P_{13}P_{23}$. Thus $\lambda_{12}=\lambda_{13}$ for all particles that are identical. The last proof noted here is that for particles with $\lambda_{ij}=-1$, the Pauli exclusion principle holds. Suppose two particles in the N-particle system, say, numbers 1 and 2, were in the same particle state. Then $\Psi(r_1,r_2,\cdots)=\Psi(r_2,r_1,\cdots)$; but $\lambda_{12}=-1$ implies that $\Psi(r_1,r_2,\cdots)=-\Psi(r_2,r_1,\cdots)$. Since the only quantity that is its own negative is zero, the wave function is identically zero for any N-particle state in which two identical particles with $\lambda_{ij}=-1$ are in the same particle state. Q.E.D.

†See Heitler, Sec. 4, or any other general quantum mechanics reference.

or atom, one simply adds the spins of the fundamental particles comprising it. Thus H, H_2, $(^4He)^{2+}$, ^4He, and D_2 are bosons and H^+, ^3He, HD, and D are fermions.

It is remarkably easy to calculate the ensemble averages of the numbers of particles in the various particle states for both fermions and bosons. These are really the ensemble averages of the occupation numbers, but most authors call them the occupation numbers of the particle states. They correspond to what one would *predict* for the number of particles in state i in a system composed of fermions or bosons.

OCCUPATION NUMBERS FOR FERMIONS AND BOSONS

For an ideal gas of fermions, each member of the canonical ensemble representing the system either will have a particle in state i or will not. The occupation number for fermions will be the same as the probability that particle state i is occupied. This is the fraction of members of the ensemble having particle state i occupied. We introduce the following notation. Let Q_N^0 and Q_N' represent the sums of terms of the form $e^{-\beta E_{\text{state}}}$ for all N-particle states in which particle state i is unoccupied or occupied, respectively. Then the ensemble average of N_i for fermions is simply the ratio of Q_N' to Q_N:

$$N_{i,FD} = \frac{Q_N'}{Q_N} = \frac{Q_N'}{Q_N^0 + Q_N'} = \frac{e^{-\beta\epsilon_i} Q_{N-1}^0}{Q_N^0 + e^{-\beta\epsilon_i} Q_{N-1}^0} \tag{9-3}$$

where we have recognized that $Q_N = Q_N^0 + Q_N'$, since all N-particle states have either zero or one particle in state i. We then factored $e^{-\beta\epsilon_i}$ out of Q_N', which left Q_{N-1}^0. Division of both the numerator and denominator of Eq. 9-3 by Q_N^0 yields

$$N_{i,FD} = \frac{e^{-\beta\epsilon_i}(Q_{N-1}^0/Q_N^0)}{1 + e^{-\beta\epsilon_i}(Q_{N-1}^0/Q_N^0)}. \tag{9-4}$$

We know from Eq. 8-20 that

$$\frac{Q_N}{Q_{N+1}} = \frac{Q_{N-1}}{Q_N} = e^{\beta\mu}, \tag{9-5}$$

where μ is the chemical potential or partial molecular Gibbs free energy. The second equality in Eq. 9-5 followed because macroscopic systems containing either N or $N-1$ particles must have essentially the same chemical potentials. We note the resemblance between the quantities in parentheses in Eq. 9-4 and the second form in Eq. 9-5. These would be identical were it not for the zero superscripts in Eq. 9-4; thus both ratios in Eq. 9-4 are $e^{\beta\mu}$ for a system like the one under consideration for which state i was unavailable to any particles. Since N different states must be occupied, whether a particular state is present or absent in the set of available states cannot significantly change the value of μ. We thus can replace the ratios in Eq. 9-4 by $e^{\beta\mu}$ and obtain the result

$$\boxed{N_{i,FD}=\frac{e^{\beta(\mu-\epsilon_i)}}{1+e^{\beta(\mu-\epsilon_i)}}} \tag{9-6}$$

for the occupation number of particle state i for the ideal gas of fermions.

For an ideal gas of bosons, each member of the ensemble will have between 0 and N particles in state i. The average number in state i is given by

$$N_{i,BE}=\frac{Q_N'+2Q_N''+3Q_N'''+\cdots}{Q}=\frac{Q_N'+2Q_N''+3Q_N'''+\cdots}{Q_N^0+Q_N'+Q_N''+\cdots}, \tag{9-7}$$

where we have extended the notation used above so that Q_N'' is the sum of terms $e^{-\beta E_{\text{state}}}$ for all N-particle states having two particles in state i, and so on. Once again, we factor $e^{-\beta\epsilon_i}$ out of each term in Eq. 9-7 as many times as it appears there and divide numerator and denominator by Q_N^0:

$$N_{i,BE}=\frac{e^{-\beta\epsilon_i}(Q_{N-1}^0/Q_N^0)+2e^{-2\beta\epsilon_i}(Q_{N-2}^0/Q_N^0)+3e^{-3\beta\epsilon_i}(Q_{N-3}^0/Q_N^0)+\cdots}{1+e^{-\beta\epsilon_i}(Q_{N-1}^0/Q_N^0)+e^{-2\beta\epsilon_i}(Q_{N-2}^0/Q_N^0)+\cdots}. \tag{9-8}$$

Next we consider each ratio in parentheses and multiply both numerator and denominator by Q_{N-s}^0 for all subscripts between that in the numerator and that in the denominator. For example, the second set of parentheses in the numerator of Eq. 9-8 is $Q_{N-2}^0/Q_N^0=(Q_{N-2}^0/Q_{N-1}^0)(Q_{N-1}^0/Q_N^0)$. The result is

$$N_{i,BE}=\frac{\left\{\begin{array}{l}e^{-\beta\epsilon_i}(Q_{N-1}^0/Q_N^0)+2e^{-2\beta\epsilon_i}(Q_{N-2}^0/Q_{N-1}^0)(Q_{N-1}^0/Q_N^0)\\ +3e^{-3\beta\epsilon_i}(Q_{N-3}^0/Q_{N-2}^0)(Q_{N-2}^0/Q_{N-1}^0)(Q_{N-1}^0/Q_N^0)+\cdots\end{array}\right\}}{1+e^{-\beta\epsilon_i}(Q_{N-1}^0/Q_N^0)+e^{-2\beta\epsilon_i}(Q_{N-2}^0/Q_{N-1}^0)(Q_{N-1}^0/Q_N^0)+\cdots}. \tag{9-9}$$

Here we note that each ratio in parentheses has the same value, so long as the series converge after a number of terms that is small compared with N. Physically, that would mean that the occupation number of all states needed be much less than N. In that case, again we can replace each ratio by $e^{\beta\mu}$, and we obtain

$$N_{i,BE}=\frac{e^{\beta(\mu-\epsilon_i)}+2e^{2\beta(\mu-\epsilon_i)}+3e^{3\beta(\mu-\epsilon_i)}+\cdots}{1+e^{\beta(\mu-\epsilon_i)}+e^{2\beta(\mu-\epsilon_i)}+\cdots}. \tag{9-10}$$

This form is immediately recognized as being*

$$\boxed{\overline{N}_{i,BE}=\frac{e^{\beta(\mu-\epsilon_i)}}{1-e^{\beta(\mu-\epsilon_i)}}.} \tag{9-13}$$

This shows that the difference between average occupation numbers for fermions and for bosons is simply the sign of the term in the denominator.

THERMODYNAMIC PROPERTIES OF FERMIONS AND BOSONS†

In this section we relate the various thermodynamic properties of ideal fermion or boson gases to the quantity ψ defined by

$$\psi\equiv\pm\sum_i \ln\left(1\pm e^{\beta(\mu-\epsilon_i)}\right), \tag{9-14}$$

*A simple proof first factors the numerator:

$$N=\frac{x+2x^2+3x^3+\cdots}{1+x+x^2+\cdots}=\frac{(x+x^2+x^3+\cdots)(1+x+x^2+\cdots)}{1+x+x^2+\cdots}, \tag{9-11}$$

which is verified directly by multiplying the two terms in the numerator.
Canceling yields

$$N=x+x^2+x^3+\cdots=\frac{x}{1-x}. \tag{9-12}$$

This can be verified by division:

$$\begin{array}{r|l} & x+x^2+\cdots \\ 1-x & x \\ & \underline{x-x^2} \\ & \quad x^2 \\ & \quad \underline{x^2-x^3} \\ & \qquad x^3 \end{array}$$

which is handy to remember.

†This section can be omitted until Chaps. 14 and 15 are studied. It is positioned here only because it fits the overall organization of the book.

where the upper (+) signs are used for fermions and the lower (−) signs are used for bosons. We view ψ as explicitly a function of β, V, and μ, and of course of the kind of particle being treated. (The influence of V is on the set of particle states i and their energies ϵ_i that arise on solution of the Schrödinger equation.)

The derivatives of ψ with respect to these variables are easily identified: For μ we write

$$\left(\frac{\partial\psi}{\partial\mu}\right)_{\beta,V}=\beta\sum_i \frac{e^{\beta(\mu-\epsilon_i)}}{1\pm e^{\beta(\mu-\epsilon_i)}}=\beta N. \tag{9-15}$$

In the last step we simply recognized the normalization condition on the occupation numbers: $\Sigma_i N_i = N$. This, in fact, is the way one would fit values of μ to different values of β, V, and N. The derivative with respect to β is

$$\left(\frac{\partial\psi}{\partial\beta}\right)_{V,\mu}=\sum_i \frac{(\mu-\epsilon_i)e^{\beta(\mu-\epsilon_i)}}{1\pm e^{\beta(\mu-\epsilon_i)}}=N\mu-E=-TS+PV. \tag{9-16}$$

The $N\mu$ term followed from recognition of N as in Eq. 9-15; the identification of $-E$ followed because the average energy is indeed $\Sigma_i N_i \epsilon_i$. The last step in Eq. 9-16 followed because $N\mu$ is the Gibbs free energy, $E-TS+PV$. The derivative with respect to V is

$$\left(\frac{\partial\psi}{\partial V}\right)_{\beta,\mu}=-\beta\sum_i \frac{(\partial\epsilon_i/\partial V)e^{\beta(\mu-\epsilon_i)}}{1\pm e^{\beta(\mu-\epsilon_i)}}=\beta P. \tag{9-17}$$

Here we utilized Eq. 6-4 in the form $-P\,dV=\Sigma_i N_i\,d\epsilon_i$, where the changes $d\epsilon_i$ in energy levels are caused by the change in volume.

Having identified the derivatives of ψ, we are in a position to identify ψ itself. The differential of the grand potential function, $\Phi\equiv E-TS-N\mu$, is given in thermodynamics by*

$$d\Phi=-S\,dT-P\,dV-N\,d\mu=kT^2S\,d\beta-P\,dV-N\,d\mu. \tag{9-18}$$

Thus the derivative of $-\beta\Phi$ is

$$d(-\beta\Phi)=-\beta\,d\Phi-\Phi\,d\beta=(-TS-\Phi)\,d\beta+\beta P\,dV+\beta N\,d\mu. \tag{9-19}$$

For large systems, thermodynamics shows Φ to equal $-PV$,* and we have the thermodynamic result

$$d(\beta PV)=(-TS+PV)\,d\beta+\beta P\,dV+\beta N\,d\mu. \tag{9-20}$$

*Andrews, pp. 83–84.

Interestingly, this is exactly the total derivative of ψ, as given by Eqs. 9-15 through 9-17. Thus βPV and ψ are within a constant of being equal. As $N \to 0$, $\beta PV \to 0$. As $N \to 0$, the only value of μ that permits the normalization, Eq. 9-15, is $\mu \to -\infty$. Thus as $N \to 0$, $\psi \to 0$ too, and we conclude that

$$\psi = \beta PV. \tag{9-21}$$

Finally, we can use Eqs. 9-16 and 9-21 to express S:

$$S = \frac{PV}{T} - \frac{1}{T}\left(\frac{\partial \psi}{\partial \beta}\right)_{V,\mu}, \tag{9-22}$$

$$S = k\psi - k\beta\left(\frac{\partial \psi}{\partial \beta}\right)_{V,\mu}. \tag{9-23}$$

Thus the thermodynamic properties of ideal fermion or boson gases are given by Eqs. 9-14 through 9-17, 9-21, and 9-23. Furthermore, for free particles in a cubical box, each translational energy level ϵ_i is proportional to $V^{2/3}$ (see Eq. 10-5). In Prob. 9-4 it is proved that this volume dependence of the energy levels implies

$$E = \frac{3}{2}PV = \frac{3\psi}{2\beta}. \qquad \text{(fermions or bosons)} \tag{9-24}$$

It is clear from the expression for the occupation number

$$N_i = \frac{e^{\beta(\mu - \epsilon_i)}}{1 \pm e^{\beta(\mu - \epsilon_i)}} \tag{9-25}$$

that N_i becomes appreciably damped only for particle energies greater than μ. The normalization condition, Eq. 9-15, is

$$N = \sum_i \frac{1}{e^{\beta(\epsilon_i - \mu)} \pm 1}. \tag{9-26}$$

For fermions, each term in the summation would have a larger denominator than for bosons with the same μ. Thus for the same value of N, μ must be larger for fermions than for bosons. Thus the damping of N_i occurs at higher energies for fermions than for bosons; thus, too, the system energy is greater for fermions than for bosons. This is not surprising. A typical measure of the energy available to a particle is kT; particle energies much larger than that are uncommon. As the temperature is lowered, the particles are squeezed into the fewer and fewer states with energy below kT. Fermions are restricted to only one particle per quantum state, whereas any number of bosons may be in the same state. Thus to accommodate all the particles, fermions must go into higher energy levels than bosons, since all the bosons may go into a single quantum state. In the extreme case of $T = 0$ K, the N fermions occupy the N states of lowest energy and all N bosons are in the single state of lowest energy. The energy (thus, the pressure, since $PV = \frac{2}{3}E$) of a fermion gas is consequently higher than it would be if the particles were bosons. These results are often said to reflect a "repulsion of the fermions due to quantum mechanical *exchange forces*" or an "attraction of the bosons leading to a quantum mechanical *condensation.*" This terminology can be misleading because we are still referring to ideal gases in which the separate particles are not directly interacting with one another.

BOLTZMANN STATISTICS

If there are a great many more available particle states than there are particles, all the occupation numbers will be much less than unity. As such, they represent the *probability* of occupancy of the particle states in the N-particle system. If either Eq. 9-6 or 9-13 is much less than unity, the $e^{\beta(\mu-\epsilon_i)}$ in the numerator must be much less than unity. That means that a good approximation to either Eq. 9-6 or 9-13, regardless of whether the system consists of fermions or bosons, is given by neglecting the $e^{\beta(\mu-\epsilon_i)}$ in the denominator (this approximation is called the **Boltzmann limit**):

$$\boxed{N_{i,FD\,\mathrm{or}\,BE} = e^{\beta(\mu-\epsilon_i)}.} \qquad \text{(Boltzmann limit)} \qquad (9\text{-}27)$$

This result, a special but as we shall see quite common case of either Fermi-Dirac or Bose-Einstein statistics, is called **Boltzmann statistics.** For systems satisfying Eq. 9-27, the probabilities of individual particle states are completely analogous to the probabilities of states of the entire system, as given by Eq. 8-30, with μ replacing A. The likelihood of a particle's being in the various possible particle states is exponentially damped by the energy of the state. This resemblance to the canonical ensemble is reasonable, since for an individual particle the rest of the particles in the system play a role similar to that taken by the heat bath in a system described by a canonical ensemble.

All the occupation numbers, given by Eq. 9-6, 9-13, or 9-27 are normalized to the total number of particles N. When Boltzmann statistics are valid, this normalization is given by

$$N=\sum_i N_i=e^{\beta\mu}\sum_i e^{-\beta\epsilon_i}=e^{\beta\mu}q. \tag{9-28}$$

Here we have defined the **particle partition function** q to be the sum of $e^{-\beta\epsilon_i}$ over all single-particle states:

$$\boxed{q\equiv\sum_i e^{-\beta\epsilon_i}}, \tag{9-29}$$

in analogy to the definition of the partition function Q for the system. Equation 9-28 may now be solved for μ:

$$\mu=kT\ln\frac{N}{q}. \tag{9-30}$$

We next find the value of Q, needed to obtain the thermodynamic properties, for the case of Boltzmann statistics. The simplest way, probably, is to combine Eq. 9-30 with Eq. 8-20 in the form

$$\mu=-kT\ln\frac{q}{N}=-kT\ln\frac{Q_{N+1}}{Q_N}, \tag{9-31}$$

$$Q_{N+1}=Q_N\frac{q}{N}=Q_{N-1}\frac{q^2}{N(N-1)}$$

$$=Q_{N-2}\frac{q^3}{N(N-1)(N-2)}=\cdots. \tag{9-32}$$

The first equality of Eq. 9-32 follows directly from Eq. 9-31; the second follows from the first on formal replacement of N by $N-1$, the third similarly on replacing $N-1$ by $N-2$, and so on. The procedure of Eq. 9-32 can be continued all the way down to Q_1, which is simply q, with the result $Q_{N+1}=q^{N+1}/N!$ or $Q_N=q^N/(N-1)!$:*

$$\boxed{Q=\frac{q^N}{N!}=\left(\frac{qe}{N}\right)^N}. \tag{9-33}$$

Here we have replaced $N-1$ by N, a procedure quite proper for huge values of N. The second equality in Eq. 9-33 is included simply for

*This procedure actually breaks down after the number becomes no longer very large. However, by that time, the overwhelming bulk of the contribution to Eq. 9-33 has been obtained.

completeness and follows from Stirling's approximation, $N! = (N/e)^N$, which is proved in most calculus books to become a more and more accurate expression for factorials as the number N becomes larger. This important result, Eq. 9-33, is valid for Boltzmann statistics only. The problem of finding the value of Q, thus the values of all the thermodynamic properties, is reduced to the calculation of the value of the particle partition function q as a function of the variables on which it depends.

A different way of viewing Eq. 9-33 perhaps makes it seem like a more obvious result. The form of q^N is that of a product of N identical factors:

$$q^N = \left(\sum_i e^{-\beta\epsilon_i}\right)\left(\sum_j e^{-\beta\epsilon_j}\right)\cdots\left(\sum_l e^{-\beta\epsilon_l}\right), \tag{9-34}$$

$$q^N = \sum_i \sum_j \cdots \sum_l e^{-\beta(\epsilon_i + \epsilon_j + \cdots + \epsilon_l)}. \tag{9-35}$$

This is similar to the form of Q:

$$Q = \sum_{\substack{N\text{-particle}\\ \text{states}}} e^{-\beta E_{\text{state}}}, \tag{9-36}$$

where an N-particle state is specified by the occupation number of each particle state, and the energy E_{state} is the sum of the energies of the occupied particle states. When Boltzmann statistics are applicable, the only significant occupation numbers are 0 and 1. Thus N different particle states are occupied in each important N-particle state. Unfortunately, Eq. 9-35 counts each set of N different particle states not once, but once for each separate *ordering* of the N states, and there are $N!$ such orderings. For example, suppose $N = 2$; the two occupied particle states r and s appear in one term in Eq. 9-36. However, in Eq. 9-35 this r–s state appears twice, once as $i = r$ and $j = s$, and again as $i = s$ and $j = r$. Suppose $N = 3$; the three particle states r, s, and t appear in one term in Q, but in q^N this r–s–t states appears six times: $i = r, j = s, k = t$; $i = r, j = t, k = s$; $i = s, j = r, k = t$; $i = s, j = t, k = r$; $i = t, j = r, k = s$; $i = t, j = s, k = r$. In general, the number of different ways in which N objects can be ordered is $N!$, and there are $N!$ equal terms in q^N that should be just one term in Q. Thus $Q = q^N/N!$.

Some authors say that a system described by Eq. 9-33 "obeys corrected Boltzmann statistics," the correction lying in the $1/N!$. These authors would say that "Boltzmann statistics" or "classical Boltzmann statistics" implies the use of q^N for Q. Such usage, however, would mean that each particle in the gas could in principle be given a label and followed as it moved around in the system. This changes the definition of a state for the N particles and changes the calculated values of the entropy, the free energies, and the chemical potential. It does not correspond to reality in a gas or liquid where the particles are free to migrate; an analogous phenomenon will be found for crystals, however, where one looks not at the constituent particles but at the different vibrational modes they generate. The author cannot think of many occasions in which q^N is used for Q, but when it is, it would make sense to refer to the system as consisting of *nonidentical* or *distinguishable* particles. Then the use of $Q = q^N/N!$ always is called a "Boltzmann gas." The molecules are identical—they are either bosons or fermions; but since many more particle states than particles are available, the difference between bosons and fermions can be neglected. The reader of the statistical mechanical literature is warned in light of the foregoing to pay attention to the definitions used by each author.

One common use of the Boltzmann distribution is in the computation of the ratio of probabilities of occupancy of two different particle states; this is identical to the ratio of occupation numbers for the states:

$$\frac{P_i}{P_j} = \frac{N_i}{N_j} = e^{-\beta(\epsilon_i - \epsilon_j)}. \tag{9-37}$$

In this case the normalization constant drops out and need not be calculated. The $e^{-\beta\epsilon_i}$ in $N_i = ae^{-\beta\epsilon_i}$ (where $a = e^{\beta\mu}$) is often called the **Boltzmann factor**. Equation 9-37 reveals why it is most unlikely for particles to be in state i if $\epsilon_i - \epsilon_j \gg kT$.

Sometimes several distinct particle states are *degenerate* (i.e., have the same energy). The **degeneracy** of an energy level is the number of different states having that same energy. Let energy level m have degeneracy g_m and level n have degeneracy g_n. Then the ratio of occupancies of the two levels is

$$\frac{N(\text{level } m)}{N(\text{level } n)} = \frac{\sum_{i=1}^{g_m} N_i(\epsilon_m)}{\sum_{j=1}^{g_n} N_j(\epsilon_n)}$$

$$= \frac{g_m N_i(\epsilon_m)}{g_n N_j(\epsilon_n)} = \frac{g_m}{g_n} e^{-\beta(\epsilon_m - \epsilon_n)}. \tag{9-38}$$

It is important to note the difference between what is expressed by Eqs. 9-37 and 9-38. The presence of the degeneracies, often called the **statistical**

weights, in Eq. 9-38 redirects one's concern away from quantum states and toward energy levels. It is useful to be able to pass between the energy level language and the quantum state language. As with Eq. 8-29, this is easily done, because the number of states contributing to a degenerate level is simply the degeneracy of the level:

$$N(\text{level of energy } \epsilon) = g(\text{energy } \epsilon) N(\text{single state of energy } \epsilon). \quad (9\text{-}39)$$

To summarize: in ideal gases there is no mechanical interaction between the particles that make up the system. The set of allowed states for an individual particle can be found by solving the Schrödinger equation for a single particle in the volume of interest. A quantum state for N particles in the volume is specified by the numbers of particles in each single-particle quantum state. The indistinguishability of like particles leads to the conclusion that all particles must be either fermions or bosons. For fermions, only the occupation numbers 0 or 1 are possible; bosons may have any number of particles in a single state. Ensemble averages of occupation numbers are easily expressed formally both for fermions and for bosons. When a great many more particle states are accessible to particles than there are particles in the system, the expressions for occupation numbers for both fermions and bosons reduce to the same form, a simple exponential analogous to that obtained for the whole system in the case of the canonical ensemble. This is called the case in which Boltzmann statistics are valid. From this it is clear that particle states are accessible if their energies ϵ_i are small compared with kT. Otherwise the $e^{-\beta\epsilon_i}$ damps their probabilities and they are not readily accessible to the particles. Thus the condition for the applicability of Boltzmann statistics is, roughly, that the number of particle states with energies below kT greatly exceeds the number of particles in the system. For such systems, $Q = q^N/N!$, where q is the particle partition function, and the task of statistical thermodynamics is reduced to that of finding q.

PROBLEMS

9-1 Consider a two-particle system (two coins), with two particle states (head, tail) of equal energy available to each. If it is impossible to tell during a collision between the two coins whether they have changed places, the coins must be either fermions or bosons. If fermions, what two-particle states are allowed? What is the probability of one head and one tail? Of two heads? If bosons, what two-particle states are allowed? What is the probability of one head and one tail? Of two heads? If the coins are treated as distinguishable, what two-particle states are allowed? What is the probability of one head and one tail? Of two heads? If Boltzmann statistics were used to describe these coins, what would be the result? Would Boltzmann statistics be useful?

9-2. Consideration of the form of Eq. 9-13 is sufficient to prove what requirement on the sign of μ for bosons?

9-3. Particle states i and j have energies -5.2×10^{-15} erg and $+1.3\times10^{-15}$ erg, respectively. At a temperature of 20 C, what is the ratio of populations of states i and j? Use Boltzmann statistics.

9-4. Prove that Eq. 9-24 results whenever $\epsilon_i \propto V^{2/3}$ for all energy levels. Equation 9-24 suggests that the zero of energy is not arbitrary, as we know it to be. What is the effect of changing the energy zero on Eq. 9-24?

9-5. Consider how the relative populations of two single-particle states i and j vary with the energy $\epsilon \equiv \epsilon_j - \epsilon_i > 0$ separating them. Sketch a plot of N_j/N_i versus ϵ for $\epsilon = 0$ up to about $\epsilon = 5kT$. For temperatures very low compared with ϵ/k, what happens to the populations of i and j? For very high temperatures compared with ϵ/k, what happens to the populations of i and j? For what temperatures, if any, does N_j exceed N_i?

9-6. Express Eqs. 9-14 through 9-17, 9-21, and 9-23 in the Boltzmann limit.

9-7. Stirling's approximation is often written in an equivalent form found by taking the logarithm of both sides of the equation given in this chapter. Derive that equivalent expression from $N! = (N/e)^N$, thus showing the origin of the e in the expression given here.

10 TRANSLATIONAL MOTION OF MOLECULES

QUANTUM MECHANICAL RESULTS

In Chap. 9 the problem of finding Q was reduced to the calculation of $(qe/N)^N = q^N/N!$ for ideal gases in which the number of particle quantum states with energies well below kT was much greater than the number of particles in the system. Thus the forbidding task of solving Schrödinger's equation for N particles is reduced to the reasonable problem of solving it for a single particle in the volume V. Since almost all books on quantum mechanics, physical chemistry, intermediate physics, or statistical mechanics contain this solution,* we will merely quote quantum mechanical results.

*In particular, see Eisberg, Sec. 8-5; Barrow, Chap. 1.

A first conclusion is that the Schrödinger equation for a single particle in the volume V separates into one equation governing the translational motion of the center of mass of the particle in the volume and another equation governing the internal motion or state of the particle. The latter equation contains contributions from rotational and vibrational motion if the particle is not a single atom, as well as from the state of the electrons. Thus the quantum state of a particle can be specified by noting first its translational state and then its internal state, the translational and internal states having no correlation with each other whatsoever. Thus q can be factored into two terms, one representing translational and the other representing internal states:

$$q=\sum_{\text{states}} e^{-\beta\epsilon_{\text{state}}}=\sum_{\substack{\text{trans}\\\text{states}}}\sum_{\substack{\text{int}\\\text{states}}} e^{-\beta[\epsilon_{\text{trans}}+\epsilon_{\text{int}}]}, \tag{10-1}$$

$$q=\left[\sum_{\substack{\text{trans}\\\text{states}}} e^{-\beta\epsilon_{\text{trans}}}\right]\left[\sum_{\substack{\text{int}\\\text{states}}} e^{-\beta\epsilon_{\text{int}}}\right] \tag{10-2}$$

$$q=q_{\text{trans}}q_{\text{int}}. \tag{10-3}$$

The notations q_{trans}, q_{int} are obvious choices for the sums over all translational and internal states of a particle.

The Schrödinger equation for the motion of the center of mass of the particle in the volume V is

$$-\frac{\hbar^2}{2m}\nabla^2\psi=\epsilon_{\text{trans}}\psi \tag{10-4}$$

with appropriate boundary condition ($\psi=0$ at walls of box). In Eq. 10-4, m is the mass of the particle; $\hbar=h/2\pi$, where h is Planck's constant; $\hbar=1.054\times10^{-27}$ erg-sec; and ψ is the wave function for the center of mass of the particle; it is a function of the position $\mathbf{r}$ of the center of mass. The wave function can satisfy Eq. 10-4 *and* be zero at the walls for only certain discrete translational quantum states, whose corresponding energies are given by

$$\epsilon_{\text{trans}}(n_x,n_y,n_z)=\frac{\pi^2\hbar^2}{2ma^2}(n_x^{\,2}+n_y^{\,2}+n_z^{\,2}); \tag{10-5}$$

n_x, n_y, n_z positive integers greater than 0;

different quantum state for each choice of n_x, n_y, *and* n_z.

For convenience, the box has been chosen as a cube of length a. Boxes with other shapes introduce no important new features. A *state* is specified by choice of the three positive integers, n_x, n_y, and n_z. These can take any integral value greater than zero, and any two or even all three of them can have the same value. Equation 10-5 shows that the translational or kinetic energy of a particle is the sum of the energy due to the x-component of its motion plus that due to the y-component of its motion plus that due to the z-component of its motion. The n's are called **translational quantum numbers.**

From time to time it is convenient to visualize each choice of n_x, n_y, and n_z as a point in the positive octant of a three-dimensional quantum number space, illustrated by Fig. 10-1. Each point in this space also can be described by a vector **n** with components n_x, n_y, n_z. Only the positive octant is used because the three quantum numbers are always positive. Equation 10-5 shows that ϵ_{trans} depends only on the magnitude of the vector **n**, and not on the angles θ and χ.

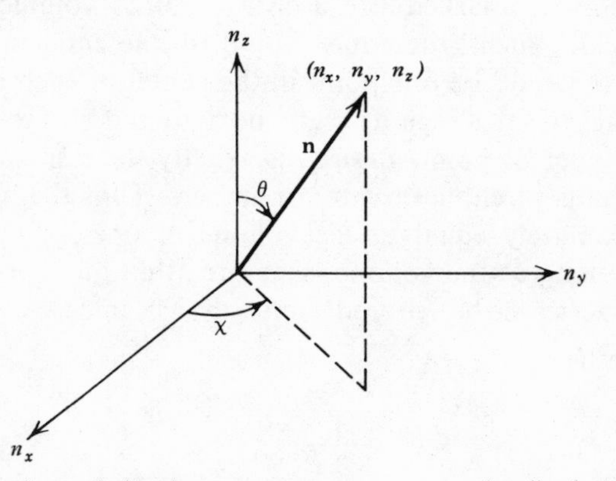

Fig. 10-1. Choice of translational quantum state, n_x, n_y, n_z, visualized as a point in three-dimensional quantum number space.

VALIDITY OF BOLTZMANN STATISTICS

To determine the kinds of system for which we would expect to be able to use Boltzmann statistics, let us calculate from Eq. 10-5 the ratio r of the number of translational particle states with energy below kT to the number of particles:

$$r \equiv \frac{\text{number of particle states with } \epsilon_i < kT}{N}. \tag{10-6}$$

A translational state will be counted for every point in the quantum number space that leads to an energy $\epsilon_{\text{trans}} < kT$. The maximum value n_{max} that the magnitude n of the vector $\mathbf{n}$ may have is easily found by equating $\epsilon_{\text{trans}}(n_{\text{max}})$ to kT:

$$\epsilon_{\text{trans}}(n_{\text{max}}) = \frac{\pi^2\hbar^2 n_{\text{max}}^2}{2ma^2} = kT, \tag{10-7}$$

$$n_{\text{max}} = \sqrt{\frac{2MkT}{N_0}}\ \frac{a}{\pi\hbar}\,. \tag{10-8}$$

Here, we have replaced the mass m of a particle by the molecular weight M divided by Avogadro's number N_0. The question is, how many points in the space of Fig. 10-1 lie within a distance n_{max} of the origin? That is, how many points lie within one-eighth of a sphere (the positive octant is only one-eighth of the sphere) whose radius is n_{max}? Imagine that about each point there is marked off a little cubical volume, one quantum number on a side. Such little cubes would fill the entire quantum number space, and there would be one point in the center of each cube. Except for some inaccuracy in the regions right next to the surfaces bounding the octant, the number of points desired is exactly the same as the volume of the octant in units of cubic quantum numbers. Thus the numerator of Eq. 10-6 is approximately equal to the volume of one-eighth of a sphere of radius n_{max}, so long as the volume is so large that the inaccuracies near its boundary surfaces can be ignored. Since the volume of a sphere of radius R is $\frac{4}{3}\pi R^3$, we have

$$r = \frac{1}{N}\,\frac{1}{8}\,\frac{4}{3}\pi n_{\text{max}}^3 \tag{10-9}$$

$$r = \frac{\pi}{6N}\left(\frac{2MkT}{N_0}\right)^{3/2} \frac{V}{\pi^3\hbar^3}\,. \tag{10-10}$$

If V/N is replaced in Eq. 10-10 by kT/P from the ideal gas law and if numerical values are inserted for all the constants, the result is (see Prob. 10-7)

$$\boxed{r = \frac{M^{3/2}T^{5/2}\ \text{atm-mole}}{50P\ \text{g}^{3/2}\text{-deg}^{5/2}}} \tag{10-11}$$

Whenever a gas has r, as given by Eq. 10-11, much greater than unity, there are enough translational states alone available to justify the use of Boltzmann statistics for a gas of either fermions or bosons. Even for a gas as light as H_2 at 1 atm pressure and only 20 K, Eq. 10-11 gives a value exceeding 50 for r. Furthermore, we have said nothing about possible internal states that might be available and would even further facilitate the applicability of Boltzmann statistics. *Thus for virtually every gas of the kind we are used to treating and for which we would be tempted to use Boltzmann statistics, that usage is fully justified.*

CONTRIBUTIONS FROM TRANSLATIONAL MOTION

We now are in a position to evaluate q_{trans} and to calculate its contributions to various thermodynamic properties. It makes the notation more convenient to consolidate all the various terms that enter in Eq. 10-5 into a single parameter. Thus we define the **characteristic temperature of translation, Θ_{trans}**:

$$k\Theta_{\text{trans}} \equiv \frac{\pi^2\hbar^2}{2ma^2} = \frac{\pi^2\hbar^2 N_0}{2MV^{2/3}}. \tag{10-12}$$

We call Θ_{trans} a temperature simply because it has the units of temperature. It is the single quantity describing both the molecules and their confining volume that is present in the solution, Eq. 10-5, of the translational Schrödinger equation. In terms of Θ_{trans}, we can write q_{trans} as

$$q_{\text{trans}} = \sum_{\substack{\text{trans}\\ \text{states}}} e^{-\epsilon_{\text{trans}}/kT}, \tag{10-13}$$

$$q_{\text{trans}} = \sum_{n_x=1}^{\infty} \sum_{n_y=1}^{\infty} \sum_{n_z=1}^{\infty} e^{-\Theta_{\text{trans}}(n_x^2+n_y^2+n_z^2)/T}, \tag{10-14}$$

$$q_{\text{trans}} = \left(\sum_{n_x=1}^{\infty} e^{-n_x^2\Theta_{\text{trans}}/T}\right)\left(\sum_{n_y=1}^{\infty} e^{-n_y^2\Theta_{\text{trans}}/T}\right) \times \left(\sum_{n_z=1}^{\infty} e^{-n_z^2\Theta_{\text{trans}}/T}\right), \tag{10-15}$$

$$q_{\text{trans}} = \left(\sum_{n=1}^{\infty} e^{-n^2\Theta_{\text{trans}}/T}\right)^3. \tag{10-16}$$

Since the quantities in the three sets of parentheses in Eq. 10-15 differ only in the label of the dummy summation index, they are identical.

Each term in the sum in Eq. 10-16 can be plotted as a dot against n, as in Fig. 10-2a. The complete sum can be viewed as the sum of the areas of an infinite number of strips, one unit wide and $e^{-n^2\Theta_{trans}/T}$ units high, as in Fig. 10-2b. The total area of all the strips could be found accurately simply by integrating the continuous function $e^{-n^2\Theta_{trans}/T}dn$, as in Fig. 10-2$c$, if the number of cases in which the height of the strip changed markedly in passing from one value of n to the next was negligible. This will be the case, provided a large number of terms are required before $e^{-n^2\Theta/T}$ is appreciably damped. We have already shown in Eq. 10-8 and 10-11 that the cube of the number of such terms is much greater than the number of particles; thus the number of such terms is greater than 10^7. Replacing the sum by the integral, therefore, leads to virtually no loss of accuracy. Finding the value of such a sum by integrating is equivalent to taking the first term of a so-called *Euler-Maclauren* expansion of the sum.*

$$q_{trans} = \left(\int_0^\infty dn\, e^{-n^2\Theta_{trans}/T} \right)^3, \tag{10-17}$$

$$\boxed{q_{trans} = \left(\frac{\pi T}{4\Theta_{trans}} \right)^{3/2} = V\left(\frac{mkT}{2\pi\hbar^2} \right)^{3/2}.} \tag{10-18}$$

The value of the integral is determined from Eq. B-1 in Appendix B, a result that is proved in Prob. 10-8.

Given q_{trans}, it is possible to calculate the translational contribution to all the thermodynamic properties using Eq. 9-33 and 10-3:

$$Q = Q_{\text{trans-indist}} Q_{int} \tag{10-19}$$

$$Q_{\text{trans-indist}} = \left(\frac{q_{trans} e}{N} \right)^N = \left(\frac{Ve}{N} \right)^N \left(\frac{mkT}{2\pi\hbar^2} \right)^{(3/2)N} \tag{10-20}$$

Into $Q_{\text{trans-indist}}$ we have lumped both the translational contribution to Q and the contribution whose origin is the indistinguishability of the particles. From this it is easy, using Eq. 8-16, to calculate the energy E_{trans} of

*Proved in most calculus books; given in Mayer and Mayer, p. 431.

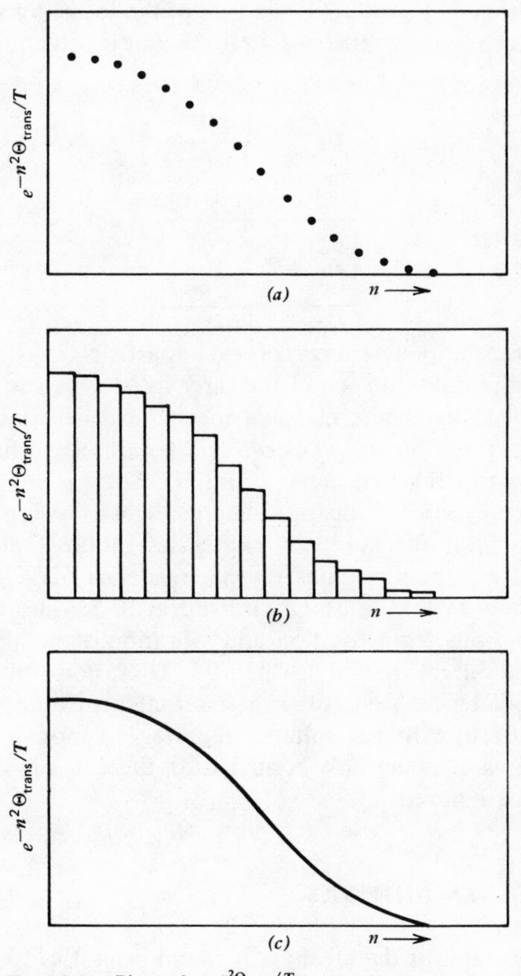

Fig. 10-2. Plots of $e^{-n^2\Theta_{\text{trans}}/T}$.

an ideal gas that is due to the translational motion of its molecules:

$$E_{\text{trans}} = kT^2 \left(\frac{\partial \ln Q_{\text{trans-indist}}}{\partial T} \right)_{N,V}, \tag{10-21}$$

$$= kT^2 \left(\frac{\partial}{\partial T} \right)_{N,V} \ln \left[\left(\frac{Ve}{N} \right)^N \left(\frac{mkT}{2\pi\hbar^2} \right)^{(3/2)N} \right]. \tag{10-22}$$

Since the logarithm of a product is the sum of the logarithms of the various factors, and since the only term in Eq. 10-22 that involves T is $T^{(3/2)N}$, the result is

$$E_{\text{trans}} = kT^2\left(\frac{\partial \ln T^{(3/2)N}}{\partial T}\right)_{N,V}, \tag{10-23}$$

$$\boxed{E_{\text{trans}} = \frac{3}{2}NkT.} \tag{10-24}$$

The average translational energy of each particle is $\frac{3}{2}kT$; or $\frac{1}{2}kT$ for motion resolved parallel to each of the three axes, x, y, and z. In Chap. 18 this is also proved using the classical mechanical description of the gas molecules. This result is the *classical limit* of the quantum mechanical problem we are treating in Chaps. 9 and 10. There is no volume dependence in the expression for ideal gas energy. This is also an experimental conclusion based on the study of real gases in the limit of vanishing density: the internal energy gains its independence of V and becomes a function of T only in that limit. Of course Eq. 10-24 does not give all the energy; contributions from rotation and vibration of molecules and from electronic excitation are treated in Chap. 11. These contributions also have no volume dependence. Regardless of internal contributions, translation contributes $\frac{3}{2}NkT$ to E for all Boltzmann gases. For monatomic gases such as the noble gases, it is the only contribution there is, provided electronic excitation can be ignored.

THE IDEAL BOLTZMANN GAS

The equation of state for the ideal gas is found using Eq. 10-20 in Eq. 8-24. Of the terms in Eq. 10-20 which comprise $Q_{\text{trans-indist}}$, the only one which depends on volume is V itself:

$$P = kT\left(\frac{\partial \ln Q}{\partial V}\right)_{T,N} = NkT\left(\frac{\partial \ln V}{\partial V}\right)_{T,N} = \frac{NkT}{V}. \tag{10-25}$$

Since the internal molecular states and energy levels do not depend on volume, Eq. 10-25 is the equation of state for *all* ideal Boltzmann gases. This is, of course, gratifying, because ever since β arose in Eq. 6-14, we have employed the result, asserted in Eq. 6-16, that $\beta = 1/kT$. With Eq. 10-25, however, this important result is finally proved. Had we been meticulous, we would have used β, rather than $1/kT$ throughout the book up to this point, where the equivalence $P = N/\beta V$ is finally proved.

The theoretical result, Eq. 10-25, for ideal Boltzmann gases is comparable to the experimental result for the pressure of real gases in the limit of vanishing density.* Many studies relating P, T, and the density N/V have been made on a wide variety of gases. In the limit of vanishing density, the ratio PV/N for all real gases at the same temperature is found to approach the same limit:

$$\lim_{N/V \to 0} \frac{PV}{N} = \frac{1}{\beta} = kT. \tag{10-26}$$

This temperature T is also found to be satisfactory for use as the integrating denominator for the reversible heat. The *scale* of temperature or size of a degree is conventionally determined by measuring the left-hand side of Eq. 10-26 for a gas in thermal equilibrium with water at its triple point. Boltzmann's constant is then fixed by defining this to be the standard temperature of 273.1600 K:

$$\lim_{N/V \to 0} \frac{PV}{N} = (273.1600\ K)k. \tag{10-27}$$

The value of $k = R/N_0$, determined from such measurements, is $(8.314 \times 10^7\ \text{erg}/K\text{-mole})/(6.022 \times 10^{23}\ \text{mole}^{-1}) = 1.3806 \times 10^{-16}\ \text{erg}/K$.

Of course, real gases never exist in the limit of vanishing density. Molecules are objects that interact with one another. It is convenient to represent the intermolecular force as the negative of the slope of a plot of intermolecular potential energy against distance of separation. In Fig. 10-3 this is done for typical pairs of helium, neon, and argon atoms. At fairly large distances of separation, on the order of 4 to 15 Å, two neutral molecules or atoms almost always attract each other. As the molecules come together, their mutual potential energy becomes smaller; the force acting on them, the slope of the potential curve, is attractive. At a certain distance of separation, the potential energy is a minimum; there is then no force between them. At closer distances, when the electron clouds are overlapping considerably, they repel each other. Their mutual potential energy increases sharply; the repulsion becomes very strong if the particles come too close together.

As the density in a gas goes down, the number of pairs of molecules close enough to interact appreciably (say, within 10 or 15 Å of each other) decreases. Finally, in the limit of vanishing density, the number of molecules within each others' fields of force is completely negligible. In that

*Andrews, pp. 47–50, Chap. 14.

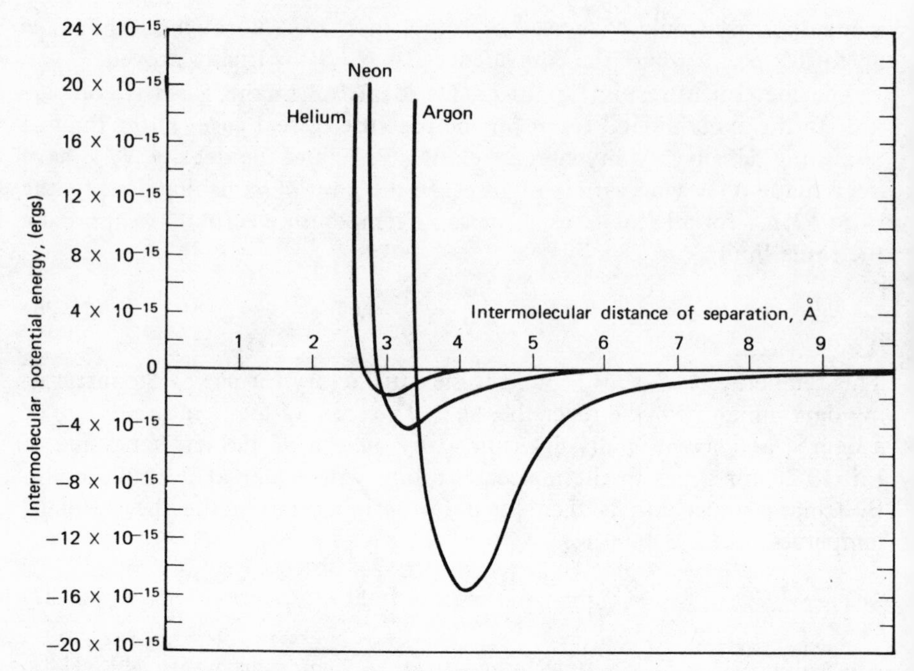

Fig. 10-3. Potential energy curves for He, Ne, and Ar. (From Hirschfelder, Curtiss, and Bird, pp. 197–199.)

limit, the molecules' interaction with each other has no bearing on the properties of the gas. The molecules are flying freely through the box; they have only translational and internal energy, no intermolecular potential energy. Since the volume affects only E_{potl}, and since temperature alone determines E_{trans} and E_{int}, ideal gases have $E = E_{\text{trans}} + E_{\text{int}}$, a quantity that is independent of V and depends only on T.

PROBLEMS

10-1. What is the ratio of populations of helium atoms in translational state $n_x = n_y = n_z = 1$ to helium atoms in translational state $n_x = n_y = 1$, $n_z = 2$? The temperature is 200 K; the volume is a cube, 10 cm on an edge.

10-2. Calculate the translation-indistinguishability contribution to the entropy. This result is called the **Sackur-Tetrode equation**. From it, derive the oft-quoted result for a monatomic gas in its standard state (ideal behavior and 1.000-atm pressure) $S^0 = \frac{3}{2}R\ln M + \frac{5}{2}R\ln T - 2.311$, in cal/K-mole.

10-3. Calculate the translation-indistinguishability contribution to the chemical potential.

10-4. Estimate the temperature below which Boltzmann statistics would begin to fail for helium gas at 1 atm pressure. How would this result be changed if r was defined more stringently to be the ratio of the number of particle states with $\epsilon_i < \frac{1}{2}kT$?

10-5. The energy levels of Eq. 10-5 are correct for a cubical box. If, however, the box is a rectangular parallelepiped of dimensions $a \times b \times c$, the levels are given by

$$\epsilon_{\text{trans}}(n_x, n_y, n_z) = \frac{\pi^2\hbar^2}{2m}\left(\frac{n_x^2}{a^2} + \frac{n_y^2}{b^2} + \frac{n_z^2}{c^2}\right). \tag{10-28}$$

Prove that q_{trans} as given by Eq. 10-18 is unchanged for the rectangular box.

10-6. Verify Eq. 10-28 by solving the Schrödinger equation. Any of the references mentioned will give considerable help.

10-7. Verify Eq. 10-11.

10-8. Prove Eq. B-1 from elementary integrals.

10-9. Find q_{trans} from Eq. 10-14, using spherical coordinates instead of Cartesian coordinates.

11 INTERNAL MOLECULAR ENERGIES

GENERAL CONSIDERATIONS

In Chap. 10, the particle partition function q was factored into the product $q_{\text{trans}}\, q_{\text{int}}$. The q_{trans}, which represents motion of the center of mass, was evaluated. This chapter considers the internal partition function q_{int}. To determine q_{int}, one must solve the Schrödinger equation for the motion of all the constituent parts of a molecule whose center of mass is fixed.

Suppose each molecule of the gas contains n atoms. Thus to determine the configuration of the molecule, $3n$ position coordinates must be fixed (e.g., the x-,y-, and z-components of each of the n atoms). Studying the mechanics of molecules is of course much easier if we do not choose those particular $3n$ coordinates as the independent variables, since no recognition that we are treating a molecule is built into them (i.e., there is no knowledge that the n atoms are bonded together). In fact, in Chap. 10 we singled out for the first three coordinates those which locate the center of mass of the molecule itself. The mechanics governing those coordinates is especially simple, and thanks to that wise choice of coordinates, an important part of the problem has been solved.

If a molecule is *linear*, the average positions of all its atoms lie on the same straight line (e.g., all diatomic molecules, CO_2, N_2O, $HC{\equiv}CH$). Such a molecule (pictured in Fig. 11-1, aligned parallel to the z-axis) can rotate in the y–z plane about the x-axis, and it can rotate in the x–z plane about the y-axis. However, rotation about the z-axis, the internuclear axis, would not involve motion of massive atomic nuclei at all. Instead, it would represent either motion of the electrons in the atoms or a spinning of the nuclei. Such motion may be considered when we treat electronic states and nuclear spins, but such motion does not involve locating the n atoms, our task at the moment. Thus a linear molecule is completely oriented by two coordinates, the angles of rotation about the two axes perpendicular to the internuclear axis.

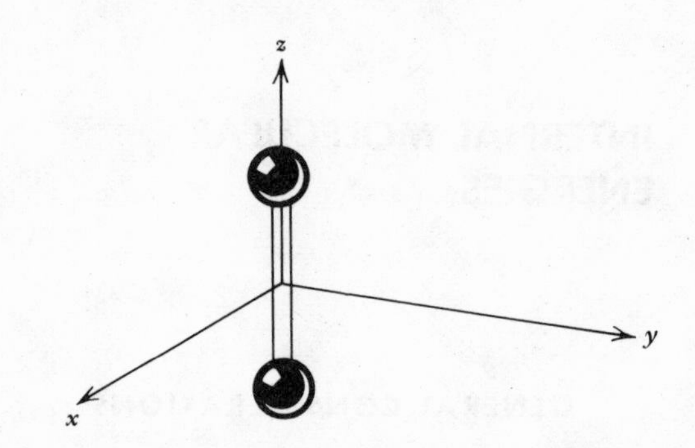

Fig. 11-1. Linear diatomic molecule, aligned with the z-axis.

If the molecule is nonlinear (e.g., H_2O, $H_2C{=}CH_2$, CH_4), the complication just described does not arise. It is necessary to specify three angles to orient such a molecule in space.

Of the $3n$ total coordinates needed to fix a molecule in space, there are now $3n-5$ left for linear molecules and $3n-6$ for nonlinear. The only freedom of motion still unspecified within the molecule is the motion of the atoms with respect to each other. For example, a diatomic molecule requires six coordinates for full description. If three of these are used to locate the center of mass and two to orient the bond axis in space, only one is left. The only unspecified item is the distance of one atom from the other; therefore, that distance must be present in the sixth coordinate chosen. It is known that the atoms in a diatomic molecule are held together by a potential energy similar to that represented in Fig. 11-2 for hydrogen. Note the resemblance between this potential and the intermolecular potentials in Fig. 10-3. Note also the important difference—that the chemical bond is some 100 to 1000 times stronger than the attractive forces between molecules—and the less important difference, that chemically bonded atoms are held more closely together than are molecules in liquids or solids. The two atoms vibrate back and forth in the potential well of Fig. 11-2. If the vibrational energy is not too great, say, up to half the energy required to dissociate the molecule, the dotted parabola is not a bad approximation to the true potential. If the parabola is used for an interatomic potential, this is called the *harmonic approximation*, because the resulting motion of the atoms is a simple harmonic oscillation.

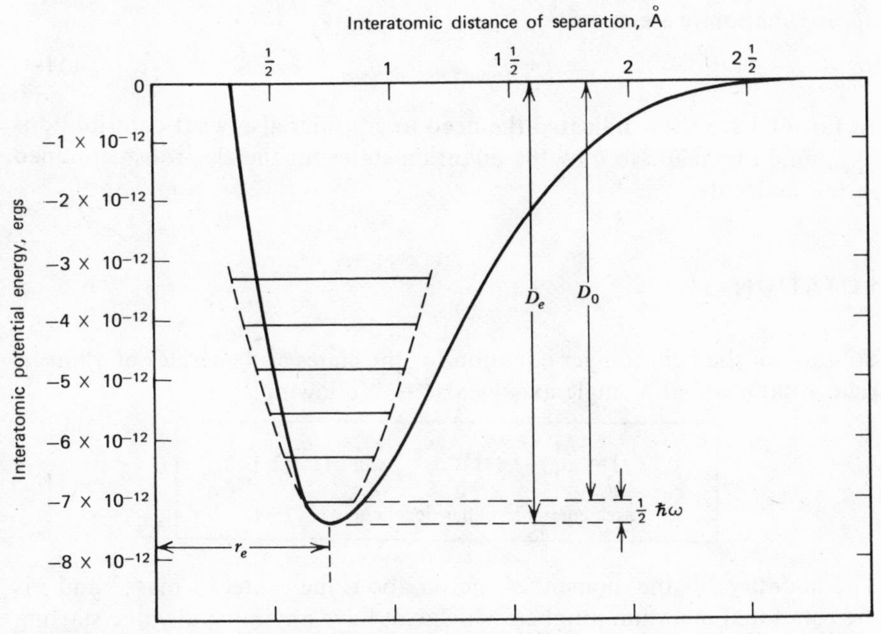

Fig. 11-2. Interatomic potential energy curve (solid line) for H_2 molecule. (From Hirschfelder, Curtiss, and Bird, p. 1058.)

For more complicated molecules with $n>2$, it is always possible to analyze the motion of the atoms into $3n-5$ or $3n-6$ so-called **normal**

coordinates. Motion along each such coordinate represents a coordinated oscillatory motion of a group of atoms. Motion along each normal coordinate is completely independent of motion along all the others. For sufficiently low energies, the harmonic approximation is quite accurate to describe the motion along each normal coordinate. Vibrational frequencies along the various coordinates need not be related to each other. Normal coordinates are further considered in Prob. 11-24.

Thus the particular $6n$ coordinates most suited to describing a molecule containing n atoms are the following: three for the molecule's center of mass, two angles to orient a linear molecule and describe its rotation, or three to orient a nonlinear molecule, and the remaining $6n-5$ or $6n-6$ normal coordinates to describe the vibrational motion of the atoms with respect to each other.

To a good first approximation, the Schrödinger equation for the internal molecular motion separates into independent equations for the rotational motion and for the vibrational motion along each normal coordinate. Corrections to this approximation are treated later in this chapter. In this approximation, we can write

$$q = q_{\text{trans}} q_{\text{rot}} q_{\text{vib}} q_{\text{elect}}. \tag{11-1}$$

In Eq. 11-1 we have indicated the need to consider also what contributions q_{elect} might be made to q by the quantum states for the electrons contained in the molecule.

ROTATION

Solution of the Schrödinger equation for the states and energies of a linear, rigid rotator about a single axis yields* the following:

$$\epsilon_{\text{rot}}(j) = \frac{\hbar^2}{2I} j(j+1); \qquad j \text{ can be } 0, 1, 2, 3, \ldots, \tag{11-2}$$

each energy level has degeneracy of $2j+1$.

The quantity I is the moment of inertia about the center of mass,† and j is the rotational quantum number, which can have any integral value starting

*All the quantum mechanics references treat this matter, but the mathematics is difficult. A simple, intuitive treatment is given by Barrow, Chap. 3.

†The relationship between the moment of inertia of a molecule and its structure is obtained from the definition of I as the sum over the atoms of the product of the atomic mass and the square of the perpendicular distance from the atom to the axis. For a diatomic molecule $I = [m_1 m_2/(m_1 + m_2)]r^2$, where r is the bond length.

with 0. Each rotational quantum state is characterized by two quantum numbers, j and m_j. There are $2j+1$ possible values of m_j for each value of j. Since choice of m_j does not affect the energy of the state, each level $\epsilon_{rot}(j)$ has a degeneracy of $2j+1$.

In Chap. 10 it was demonstrated that a quantum of translational energy—namely, the energy needed to excite a typical molecule into a higher translational state—was far less than kT for any reasonable temperature. This was equivalent to the demonstration that Θ_{trans} was far less than any reasonable T. With rotational quantum states, we again are interested in knowing the size of a quantum of energy, and again it is convenient to lodge the information about the rotation which enters Eq. 11-2 (i.e., the value of I) in a parameter Θ_{rot}, the so-called **characteristic temperature of rotation**:

$$k\Theta_{rot} \equiv \frac{\hbar^2}{2I}. \tag{11-3}$$

Values of Θ_{rot}, along with the corresponding moments of inertia from which they are calculated, are given in Table 11-1.

Table 11-1. Parameters for Common Diatomic Molecules. Source: Based on a table in Hill, p. 153.

	Θ_{rot}, K	I, g-cm^2	Θ_{vib}, K	ω, rad/sec	D_0, erg	r_e, Å
H_2	85.4	0.0471×10^{-39}	6210	8.13×10^{14}	7.14×10^{-12}	0.740
HCl	15.2	0.265×10^{-39}	4140	5.42×10^{14}	7.09×10^{-12}	1.275
HBr	12.1	0.333×10^{-39}	3700	4.84×10^{14}	5.77×10^{-12}	1.414
HI	9.0	0.45×10^{-39}	3200	4.19×10^{14}	4.41×10^{-12}	1.604
N_2	2.86	1.41×10^{-39}	3340	4.37×10^{14}	11.82×10^{-12}	1.095
CO	2.77	1.45×10^{-39}	3070	4.02×10^{14}	14.64×10^{-12}	1.128
NO	2.42	1.66×10^{-39}	2690	3.52×10^{14}	8.48×10^{-12}	1.150
O_2	2.07	1.94×10^{-39}	2230	2.92×10^{14}	8.14×10^{-12}	1.204
Cl_2	0.348	11.6×10^{-39}	810	1.06×10^{14}	3.97×10^{-12}	1.989
Br_2	0.116	34.7×10^{-39}	470	0.62×10^{14}	3.16×10^{-12}	2.284
I_2	0.054	$75. \times 10^{-39}$	310	0.41×10^{14}	2.47×10^{-12}	2.667

We now calculate q_{rot} for linear molecules, which of course is the class to which all diatomic molecules belong:

$$q_{rot} = \sum_{\substack{\text{rot}\\ \text{states}}} e^{-\epsilon_{rot}/kT} = \sum_{\substack{\text{rot}\\ \text{states}}} e^{-j(j+1)\Theta_{rot}/T}. \tag{11-5}$$

There are two kinds of linear molecule, those that are symmetric about the center of mass (e.g., O_2, N_2, HC=CH) and those that are asymmetric (e.g., HCl, CO, N_2O). The indistinguishability of like particles places no quantum restriction on rotational quantum states for asymmetric molecules; *asymmetric linear molecules may have any value of j*. However, the proper quantum treatment of the indistinguishability of like particles leads to the rule that *symmetric linear molecules are restricted either to only odd values of j or to only even values of j*, depending on the spins of the nuclei. Every energy level for an allowed value of j always has its degeneracy of $2j+1$.

It is clear from Table 11-1 that most temperatures one will be interested in will lie much greater than Θ_{rot}. For such temperatures, changing from one value of j to the next will not change the value of $\exp[-j(j+1)\Theta_{rot}/T]$ very much. The same argument used in Chap. 10 would also permit here the evaluation of the sum in Eq. 11-5 by integrating, except for the restriction to odd or even states for symmetric molecules. For such molecules, however, counting only half the states will lead to a value for q_{rot} just half as big as would be found if all the states were counted. Thus we first integrate to obtain q_{rot} as if all states were counted. We then divide the result by the so-called **symmetry number** σ, which is 1 for an asymmetric molecule and 2 for a symmetric linear molecule:*

$$q_{rot} = \frac{1}{\sigma} \sum_{j=0}^{\infty} (2j+1) e^{-j(j+1)\Theta_{rot}/T}. \tag{11-6}$$

Evaluating the sum by integration, for $T \gg \Theta_{rot}$, yields

$$q_{rot} = \frac{1}{\sigma} \int_0^{\infty} dj\,(2j+1)\,e^{-j(j+1)\Theta_{rot}/T}. \tag{11-7}$$

The integration is performed remarkably easily by making the following change of variable:

$$z = j(j+1); \qquad dz = (2j+1)\,dj, \tag{11-8}$$

$$q_{rot} = \frac{1}{\sigma} \int_0^{\infty} dz e^{-z\Theta_{rot}/T}, \tag{11-9}$$

$$\boxed{q_{rot} = \frac{T}{\sigma\Theta_{rot}} = \frac{2IkT}{\sigma\hbar^2}.} \qquad \text{(linear molecule)} \tag{11-10}$$

*Further discussed later in this chapter. Extensively treated by Mayer and Mayer, Sec. 7-b.

Thus the rotational contribution to q proves to be surprisingly simple.

From Eq. 11-10 we now calculate the rotational contribution to the energy for linear molecules:

$$E_{\text{rot}} = kT^2 \frac{\partial}{\partial T} \ln q_{\text{rot}}^N = NkT^2 \frac{\partial \ln q_{\text{rot}}}{\partial T}, \tag{11-11}$$

$$\boxed{E_{\text{rot}} = NkT.} \qquad \text{(linear molecule)} \tag{11-12}$$

This is the classical limit ($T \gg \Theta_{\text{rot}}$) of the energy contributed by the 2 rotational degrees of freedom in the linear molecule. The contribution, $\frac{1}{2}kT$ per degree of freedom per molecule, is exactly the same as that previously obtained for each translational degree of freedom.

Of course, if the temperature is not high enough to warrant integrating to find the value of Eq. 11-6, a more complicated result is obtained.* For $T \ll \Theta_{\text{rot}}$, almost all the molecules are constrained by the Boltzmann factor to be in the lowest possible rotational state, $j=0$. The energy to excite molecules rotationally must come from collisions between molecules in the gas. At these low temperatures, colliding molecules usually have so little kinetic energy that it is impossible for collisions to furnish a whole quantum of rotational energy. Thus for temperatures $T \ll \Theta_{\text{rot}}$, E_{rot} is negligible; one might say the rotational energy is "frozen out." Then, as T increases to become comparable to Θ_{rot}, more of the molecular collisions are energetic enough to leave one or more of the colliding particles rotationally excited. The Boltzmann factor is less severely damping, and excited rotational states assume reasonable probabilities. The value of E_{rot} gradually increases until it approaches its classical value of NkT. The rotational degrees of freedom "thaw out" at temperatures around Θ_{rot}. This is pictured in Fig. 11-3*a*.

The slope of the energy curve is the rotational contribution to the specific heat C_V. Since almost no energy is present in rotational contributions at low temperatures, $C_{V,\text{rot}}$ is very small for $T \ll \Theta_{\text{rot}}$. As the rotational degrees of freedom thaw out and the molecules begin to rotate, $C_{V,\text{rot}}$ increases. It must go through a maximum as E_{rot} hurries to catch up with the classical result NkT. Beyond that, $C_{V,\text{rot}}$ slowly approaches its asymptotic value Nk (see Fig. 11-3*b*).

In the classical limit where $T \gg \Theta_{\text{rot}}$, q_{rot} for nonlinear molecules is given

*Mayer and Mayer, pp. 151–157.

by a straightforward extension of Eq. 11-10:*

$$q_{\text{rot}} = \frac{\pi^{1/2}}{\sigma}\left(\frac{2I_x kT}{\hbar^2}\right)^{1/2}\left(\frac{2I_y kT}{\hbar^2}\right)^{1/2}\left(\frac{2I_z kT}{\hbar^2}\right)^{1/2}, \qquad \text{(nonlinear molecule)} \tag{11-13}$$

where I_x, I_y, and I_z are the three principal moments of inertia of the nonlinear molecule. The symmetry number σ for a nonlinear molecule is defined and discussed at the end of this chapter. In Prob. 11-9, it is proved that the contribution of Eq. 11-13 adds the following to the energy and the

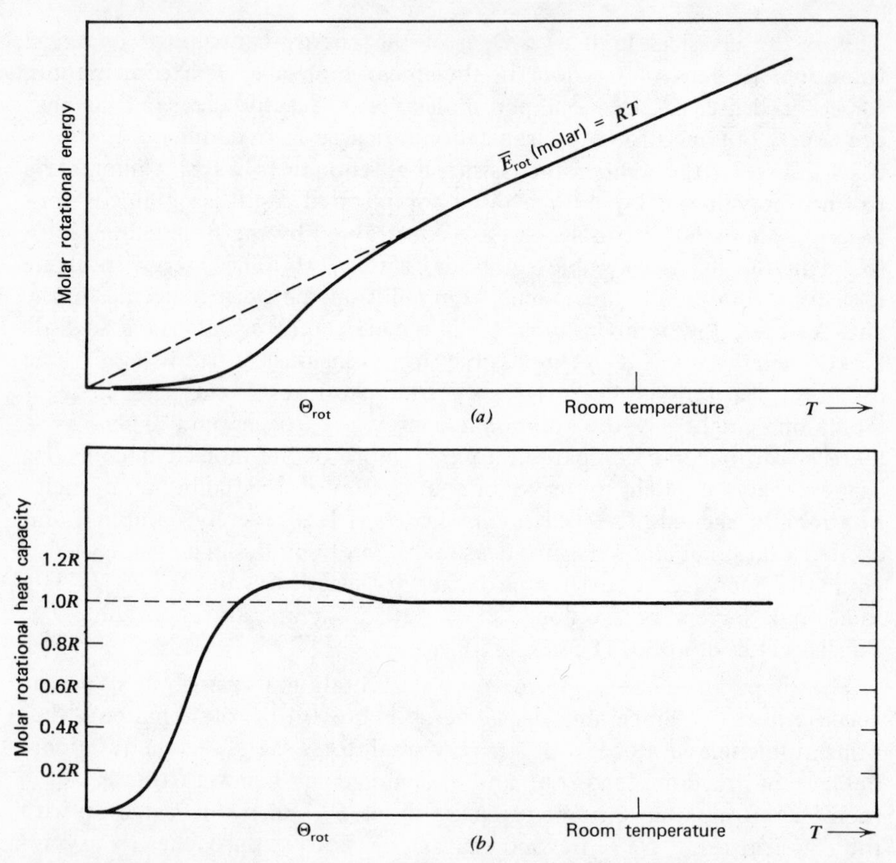

Fig. 11-3. Rotational contributions to (*a*) molar energy and (*b*) molar heat capacity of ideal diatomic gas as function of temperature.

*Mayer and Mayer, pp. 191–199.

heat capacity:

$$E_{\text{rot}} = \tfrac{3}{2}NkT; \qquad C_{V,\text{rot}} = \tfrac{3}{2}Nk. \qquad \text{(nonlinear molecule)} \quad (11\text{-}14)$$

These are the classical values, an average of $\frac{1}{2}kT$ of rotational energy about each axis of rotation for each molecule.

VIBRATION

The only vibrational motion we consider is that of the two atoms of a diatomic molecule about its center of mass. Since the normal coordinates of a polyatomic molecule resemble independent vibrations, this discussion will give adequate background for the understanding of more complicated systems.

First, we summarize briefly the classical mechanics of vibrational motion. Suppose two particles interact with each other through the parabolic potential given by the dashed line of Fig. 11-2. Let r be the distance separating the two particles, and r_e their so-called equilibrium distance, (i.e., the value of r when $U(r)$ is a minimum). Then define a new variable

$$x \equiv r - r_e \qquad (11\text{-}15)$$

to be the value of r relative to r_e. A plot of U versus x is given in Fig. 11-4 for a parabolic potential

$$U = \tfrac{1}{2}kx^2. \qquad (11\text{-}16)$$

The quantity k is called the **force constant** for the vibration; the larger the value of k, the steeper sloping are the sides of the potential well, and the narrower the bottom of the well. The classical vibrational motion of the

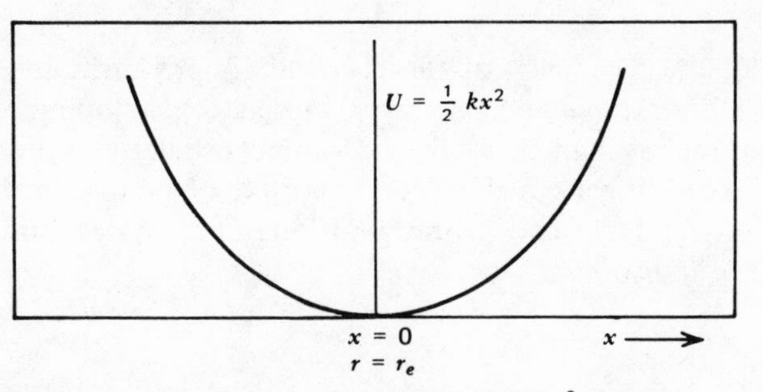

Fig. 11-4. Plot of parabolic potential $U = \frac{1}{2}kx^2$ versus x.

two atoms in a diatomic molecule is represented by the motion of a single particle of mass

$$\mu = \frac{m_1 m_2}{m_1 + m_2} \tag{11-17}$$

in the well of Fig. 11-4.* The quantity μ is called the **reduced mass** of the molecule. The force exerted on the particle of mass μ at any position x is the negative slope of the potential energy curve:

$$F = -\frac{dU}{dx} = -kx = \mu \frac{d^2x}{dt^2}, \tag{11-18}$$

where the last identification follows from Newton's law, $F = ma$. The solution of the differential equation

$$\ddot{x} = -\left(\frac{k}{\mu}\right)x \tag{11-19}$$

given by Eq. 11-18 is sinusoidal motion of the form

$$x = A\cos(\omega t) \tag{11-20}$$

where ω is the angular velocity of the motion (rad/sec). The circular frequency ν (cycles/sec) is related to ω by $\nu = \omega/2\pi$. That Eq. 11-20 indeed solves Eq. 11-19 is shown immediately by differentiating Eq. 11-20 twice:

$$\dot{x} = -\omega A \sin(\omega t); \qquad \ddot{x} = -\omega^2 A \cos(\omega t) = -\omega^2 x. \tag{11-21}$$

Thus Eq. 11-21 is identical to Eq. 11-19, and the frequency must be related to the force constant by

$$\omega = \sqrt{\frac{k}{\mu}}\ . \tag{11-22}$$

We note that the frequency of the classical simple harmonic vibration is greater for lighter atoms and narrower, more steeply sloping potentials.

The quantum mechanics of the harmonic oscillator is determined by solving the Schrödinger equation for a particle of mass μ in the parabolic potential, Eq. 11-16.† The vibrational quantum states and associated energies are found to be

*Barrow, pp. 25–34.

†All quantum mechanics books treat this subject. For example, Barrow, Chap. 2; Eisberg, Sec. 8-6.

$$\epsilon_{vib}(n) = (n + \tfrac{1}{2})\hbar\omega \equiv (n + \tfrac{1}{2})h\nu; \qquad n \text{ may be } 0, 1, 2, 3, \cdots;$$
$$\text{no degeneracy arises from vibration.} \tag{11-23}$$

The vibrational quantum number n may have any integral value from zero to infinity. No degeneracies are introduced by vibrational energy levels. In a manner similar to that found in Eq. 10-5 for the translation of a particle in a box, no mutually vibrating pair of atoms can have zero energy. If such a pair did, the atoms would have a definite location with respect to each other and a definite velocity (zero). This would violate the Heisenberg uncertainty principle. The energy levels of Eq. 11-23 appear as horizontal lines in the parabola of Fig. 11-2. The lowest vibrational state lies $\frac{1}{2}\hbar\omega$ above the bottom of the potential curve. This energy is often referred to as the **zero point vibrational energy**. It represents vibrational motion of the molecule in its ground state. Use of Eq. 11-16 and Eq. 11-23 implies that we have chosen the bottom or lowest point on the potential energy curve to be the zero of energy. Since $\frac{1}{2}\hbar\omega$ is as low as the energy of the molecule can be, we could have chosen that to be the energy zero, thus eliminating zero-point energies from consideration. Means of changing at will from one choice of energy zero to another are discussed later in this chapter.

In Table 11-1, vibrational data are tabulated for common molecules in terms of both the angular velocity ω and the **characteristic temperature of vibration** Θ_{vib}, defined by

$$k\Theta_{vib} \equiv \hbar\omega \equiv h\nu. \tag{11-24}$$

Also given in Table 11-1 are values of the depths of the vibrational ground states below the energy at infinite separation. These depths represent the energy needed to separate the two atoms when the molecule is in its ground state; that is, the values are the maximum strengths of the chemical bonds in the molecule. Table 11-1 also shows the equilibrium separation distance r_e.

Note: Information on vibrational energies is rarely given in terms of angular velocity ω in radians per second. It is more commonly given in terms of vibrational frequency $\nu = \omega/2\pi$ in reciprocal seconds. It is also given sometimes in energy units such as ergs, meaning the value of a vibrational quantum, $\hbar\omega = h\nu$. Perhaps the most common form is in units of wave number (cm^{-1}), meaning the value $\bar{\nu} \equiv 1/\lambda = \nu/c$ corresponding to a photon of energy equal to the vibrational quantum.

Because at room temperatures most molecules have Θ_{vib} large compared with T, most gas molecules are vibrationally unexcited. The quantum of vibrational energy, $\hbar\omega$, is large compared with kT. This is convenient, because the parabolic potential is a poor approximation to the real potential, especially for energies higher than about $\frac{1}{2}D_0$. So few molecules at reasonable temperatures have this much vibrational energy that this difficulty usually can be neglected and Eq. 11-23 used.

On the other hand, it is of course unreasonable to integrate to find q_{vib}:

$$q_{vib} = \sum_{n=0}^{\infty} e^{-(n+1/2)\Theta_{vib}/T}, \tag{11-25}$$

$$q_{vib} = e^{-\Theta_{vib}/2T} \sum_{n=0}^{\infty} \left(e^{-\Theta_{vib}/T}\right)^n. \tag{11-26}$$

The rewriting of Eq. 11-26 suggests a way to sum the series. In Eq. 9-12 we have already used a series analogous to the one we want, namely,

$$\sum_{n=0}^{\infty} x^n = \frac{1}{1-x}, \tag{11-27}$$

which can be verified immediately by dividing out the right-hand side. Thus Eq. 11-26 becomes

$$\boxed{q_{vib} = \frac{e^{-\Theta_{vib}/2T}}{1-e^{-\Theta_{vib}/T}}.} \tag{11-28}$$

The numerator of Eq. 11-28 represents the zero-point energy. The exponential in the denominator represents the contribution of vibrationally excited molecules.

Knowing q_{vib}, it is easy to find E_{vib}:

$$E_{vib} = NkT^2\left(\frac{\partial \ln q_{vib}}{\partial T}\right)_{V,N} \tag{11-29}$$

$$= \tfrac{1}{2}Nk\Theta_{vib} + \frac{Nk\Theta_{vib}e^{-\Theta_{vib}/T}}{1-e^{-\Theta_{vib}/T}} \tag{11-30}$$

$$E_{vib} = \tfrac{1}{2}Nk\Theta_{vib} + \frac{Nk\Theta_{vib}}{e^{\Theta_{vib}/T}-1}. \tag{11-31}$$

The first term in Eq. 11-31 is the zero-point energy. The value of Eq. 11-31 is determined relative to an energy zero chosen for each diatomic molecule as the *bottom* of its intermolecular potential well. A different choice of zero would add to Eq. 11-31 the energy of the bottom of the potential wells relative to the new energy zero. The vibrational contribution to C_V is

$$C_{V,\text{vib}} = \left(\frac{\partial E_{\text{vib}}}{\partial T}\right)_V = \frac{Nk\Theta_{\text{vib}}^2 e^{\Theta_{\text{vib}}/T}}{T^2(e^{\Theta_{\text{vib}}/T}-1)^2}. \tag{11-32}$$

In Fig. 11-5, the vibrational contributions to the molar energy and heat capacity of an ideal diatomic gas are plotted as functions of temperature.

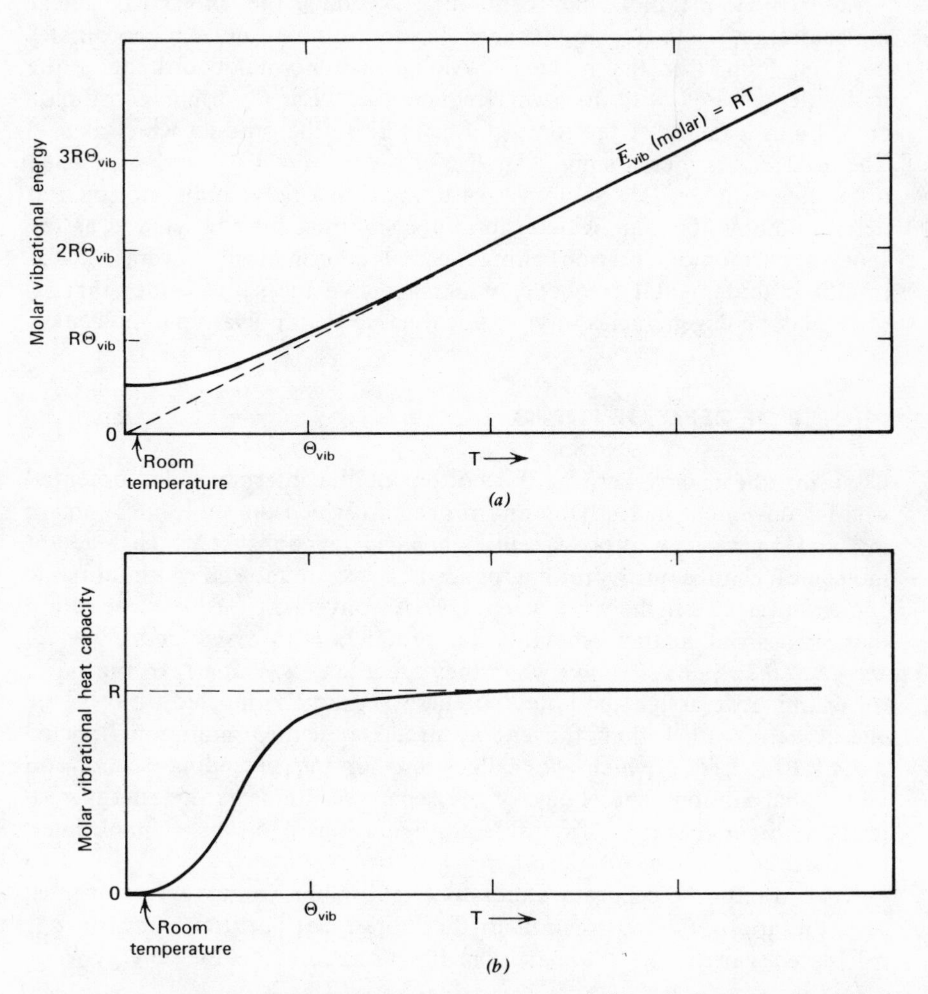

Fig. 11-5. Vibrational contributions to (*a*) molar energy and (*b*) molar heat capacity of ideal diatomic gas as function of temperature.

Vibrational modes thaw out over a relatively much greater temperature range than rotational modes. Also, there is no hump in the heat capacity curve because E_{vib} starts above its asymptotic line at $T=0$, rather than below it as in the case of E_{rot}. In Prob. 11-10 it is verified that the high-temperature limit of E_{vib} is NkT, which is shown in Chap. 18 to be the classical limit for vibrational energy. Each simple harmonic oscillator contributes kT to E_{vib}, rather than $\frac{1}{2}kT$. The reason, presented in Chap. 18, is that vibrational motion involves both kinetic and potential energy, each one contributing an average of $\frac{1}{2}kT$.

For polyatomic molecules containing n atoms, the $3n-5$ (for linear molecules) or $3n-6$ (for nonlinear molecules) different normal coordinates must be found for the molecule. Along each normal coordinate i the molecule vibrates with its own frequency ω_i. Each normal coordinate contributes a factor of the form of Eq. 11-28 to the product which is q_{vib}. The total E_{vib} is then a sum of an E_{vib} of the form of Eq. 11-31 from each normal coordinate. It is seldom easy to analyze a polyatomic molecule to determine how its complicated vibrational motions may be viewed as the sum of harmonic vibration along normal coordinates.* Several of the resulting fundamental frequencies ω_i may have the same value; that is, there may be degeneracies in vibrational energies of polyatomic molecules.

CHOICE OF ZERO OF ENERGY

Choosing the energy zero at the bottom of the intermolecular potential well is convenient in studying any process in which the molecules remain intact. However, for processes like chemical reactions in which reactant molecules change into product molecules, which molecules should one choose to establish the zero of energy? A convenient means of selection that singles out neither reactants nor products is to *assign zero potential energy to the separated atoms* when they are a long way apart. In Fig. 11-2, for example we assign the potential energy at large r (labeled 0) to be the energy zero, rather than the energy at the potential minimum (labeled -7.4×10^{-12} erg), which we called zero in the preceding section on vibrational motion. The choice of the separated atoms as the energy zero leads to *negative energy* for stable configurations of atoms in molecules, but that should cause no alarm.

If we do choose separate atoms to establish the energy zero, we must seek the appropriate correction to the vibrational partition function q_{vib} calculated earlier. In the partition function, $\Sigma_i e^{-\beta\epsilon_i}$, the energy ϵ_i is

*Barrow, Chap. 6; Karplus and Porter, Sec. 7.9.

determined relative to whatever zero has been chosen. The energy of state i with respect to separated atoms is of course just the difference between ϵ_i and the energy of the separated atoms:

$$\epsilon_i(\text{atoms zero}) = \epsilon_i(\text{arbitrary zero}) - \epsilon_{\text{separate atoms}}(\text{arbitrary zero}). \quad (11\text{-}33)$$

The right-hand side of Eq. 11-33 is independent of choice of the arbitrary zero, so long as the same choice is used in both terms. We now can write q_{vib} (atoms zero) in terms of q_{vib} calculated from any other zero of energy. We start with the definition of q_{vib} (atoms zero) as the sum over states:

$$q_{\text{vib}}(\text{atoms zero}) = \sum_i e^{-\beta\epsilon_i(\text{atoms zero})} \quad (11\text{-}34)$$

$$= e^{\beta\epsilon_{\text{separate atoms}}(\text{arbitrary zero})} \sum_i e^{-\beta\epsilon_i(\text{arbitrary zero})}. \quad (11\text{-}35)$$

In Eq. 11-35 we have simply inserted Eq. 11-33 into Eq. 11-34. The arbitrary zero in which we are interested here is the choice used above of the bottom of the intermolecular potential well. For that choice, the summation in Eq. 11-35 is the q_{vib} of Eq. 11-28, and (see Fig. 11-2)

$$\epsilon_{\text{separate atoms}}(\text{well bottom}) = D_e = D_0 + \tfrac{1}{2}\hbar\omega = D_0 + \tfrac{1}{2}k\Theta_{\text{vib}}. \quad (11\text{-}36)$$

Thus we conclude that

$$q_{\text{vib}}(\text{atoms zero}) = e^{\beta D_e} q_{\text{vib}} = e^{\beta(D_0 + (1/2)\hbar\omega)} q_{\text{vib}} \quad (11\text{-}37)$$

$$= e^{\beta(D_0 + (1/2)k\Theta_{\text{vib}})} q_{\text{vib}} \equiv q_{\text{zero}} q_{\text{vib}}, \quad (11\text{-}38)$$

where we have defined q_{zero} as $e^{\beta D_e}$, a factor correcting q_{vib} for the change in zero. The contribution E_{zero} of q_{zero} to the energy is of course

$$E_{\text{zero}} = -ND_e = -ND_0 - \tfrac{1}{2}N\hbar\omega = -ND_0 - \tfrac{1}{2}Nk\Theta_{\text{vib}}, \quad (11\text{-}39)$$

as is proved in Prob. 11-16.

ELECTRONIC STATES

We have so far ignored the electronic states of the atom or molecule being considered. In principle, the Schrödinger equation is solved for the electrons in the region of the positive nuclei.* The result is a set of electronic quantum states and their respective energies. This information is also obtainable from spectroscopic experiments. The electronic partition function for an atom or molecule is

$$q_{\text{elect}} = \sum_{\substack{\text{elect}\\ \text{states}}} e^{-\beta\epsilon_{\text{state}}} \tag{11-40}$$

Quite commonly, the excited electronic levels of the atoms or molecules of interest lie so far above the ground state that only at temperatures of several thousand degrees do they thaw out appreciably and assume significant probability. Thus only the ground electronic state, of degeneracy g_0, contributes to the sum in Eq. 11-39:

$$q_{\text{elect}} \approx \sum_{\substack{\text{states of}\\ \text{zero energy}}} e^0 = \sum_{\substack{\text{states of}\\ \text{zero energy}}} 1, \tag{11-41}$$

$$\boxed{q_{\text{elect}} = g_0.} \tag{11-42}$$

Here we have chosen the electronic ground state to be the zero of energy.

The Cl atom, for example, has a ground state energy level with degeneracy of 4. Thus q_{elect} for a gas of Cl atoms is 4, and the electronic contribution to ln Q is $N \ln 4$. This adds $Nk \ln 4$ to the entropy of the system, although it does not change the energy or the heat capacity. The NO molecule, on the other hand, has a doubly degenerate ground state and a doubly degenerate low-lying excited level of energy $(178\,\text{K})k$ above the ground state. Thus for NO, the excited level is appreciably populated, and it contributes appreciably to E and C_V. This is computed in Prob. 11-11.

SUMMARY

The statistical thermodynamics of ideal monatomic or diatomic gases, described in the last two chapters, is surprisingly simple. It is summarized in Table 11-2. When quantum levels for a given contribution are widely spaced with respect to kT, most molecules are in the lowest level. The contribution to E and C_V at such low temperatures is negligible. As the

*Barrow, Chaps. 10 and 11; Karplus and Porter, Chaps. 5 and 6.

temperature is raised, the various contributions thaw out. Even at the very lowest temperatures, the translational motion of the centers of mass of the atoms or molecules is essentially classical; translational energy levels are infinitesimally separated compared with kT. Rotational motion thaws out at very low temperatures, and long before room temperature is reached, rotational energy has become classical. The vibrational levels thaw out only at temperatures from a few hundred to a few thousand kelvin. Finally, electronic levels can be excited at high enough temperatures and can contribute in complicated ways to E and C_V.

Thus the heat capacity for an ideal gas at very low temperatures ($T \ll \Theta_{rot}$) is just $\frac{3}{2}Nk$. The only sink for energy in the process of raising the temperature of the gas is the translation of the molecules. The rest of the available levels are inaccessible because they require much more energy than the typically available kT to become populated. As the temperature is increased ($T > \Theta_{rot}$), C_V rises to $\frac{5}{2}Nk$ for linear molecules or to $\frac{6}{2}Nk$ for nonlinear ones. Now to raise the temperature by one degree, energy must be put into both translation and rotation. Then, as the vibrational levels thaw out (at $T \approx \Theta_{vib}$), C_V rises until it reaches $(3n - \frac{5}{2})Nk$ for linear molecules or $(3n-3)Nk$ for nonlinear ones.

VIBRATIONAL AND ROTATIONAL COMPLICATIONS

Using the foregoing simple results for q_{rot} and q_{vib}, we ignore a number of complications. As Fig. 11-2 indicates, approximating the interatomic potential energy curve along a chemical bond axis by a parabola may be reasonable for the lowest vibrational levels, but the anharmonic nature of the actual potential will introduce distortions. This is especially true at high temperatures of the order of Θ_{vib}. Furthermore, Eq. 11-2 implies that the moment of inertia of the molecule is constant, independent of how fast the molecule is rotating or how hard it is vibrating. Yet, clearly, rotation causes centrifugal forces that will stretch the bond, thus increasing the value of I. Also, since I will change throughout the course of a single vibration, the I of Eq. 11-2 must be the average of the instantaneous value taken over a whole period of vibration. Because the interatomic potential is anharmonic, the average interatomic separation increases with vibrational energy, and I therefore increases with vibrational energy as well as with rotational.

Corrections* for these various features can be made for a diatomic

*Discussions of Eq. 11-43 and the effect it makes on the partition function are given by Mayer and Mayer, pp. 160–166, and by Davidson, pp. 116–119.

Table 11-2. Contributions to Thermodynamic Variables in an Ideal Diatomic Gas.

$$Q=\left(\frac{e}{N}q\right)^N=\left(\frac{e}{N}q_{\text{trans}}q_{\text{rot}}q_{\text{vib}}q_{\text{zero}}q_{\text{elect}}\right)^N$$

	Translation-Indistinguishability	Rotation	Vibration	Zero of Energy	Electronic
q	$\frac{e}{N}q_{\text{trans}}=\frac{eV}{N}\left(\frac{mkT}{2\pi\hbar^2}\right)^{3/2}$	$q_{\text{rot}}=\frac{T}{\sigma\Theta_{\text{rot}}}$	$q_{\text{vib}}=\frac{e^{-\Theta_{\text{vib}}/2T}}{1-e^{-\Theta_{\text{vib}}/T}}$	$q_{\text{zero}}=e^{\beta D_e}$	$q_{\text{elect}}=g_0$
E	$\frac{3}{2}NkT$	NkT	$\frac{1}{2}Nk\Theta_{\text{vib}}+\frac{Nk\Theta_{\text{vib}}}{e^{\Theta_{\text{vib}}/T}-1}$	$-ND_0-\frac{1}{2}Nk\Theta_{\text{vib}}$	0
A	$-NkT\ln\left[\frac{eV}{N}\left(\frac{mkT}{2\pi\hbar^2}\right)^{3/2}\right]$	$-NkT\ln\left(\frac{T}{\sigma\Theta_{\text{rot}}}\right)$	$\frac{1}{2}Nk\Theta_{\text{vib}}+NkT\ln(1-e^{-\Theta_{\text{vib}}/T})$	$-ND_0-\frac{1}{2}Nk\Theta_{\text{vib}}$	$-NkT\ln g_0$
μ_i	$-kT\ln\left[\frac{V}{N}\left(\frac{mkT}{2\pi\hbar^2}\right)^{3/2}\right]$	$-kT\ln\left(\frac{T}{\sigma\Theta_{\text{rot}}}\right)$	$\frac{1}{2}k\Theta_{\text{vib}}+kT\ln(1-e^{-\Theta_{\text{vib}}/T})$	$-D_0-\frac{1}{2}k\Theta_{\text{vib}}$	$-kT\ln g_0$
S	$Nk\ln\left[\frac{e^{5/2}V}{N}\left(\frac{mkT}{2\pi\hbar^2}\right)^{3/2}\right]$	$Nk\ln\left(\frac{eT}{\sigma\Theta_{\text{rot}}}\right)$	$-Nk\ln(1-e^{-\Theta_{\text{vib}}/T})+\frac{Nk\Theta_{\text{vib}}}{T(e^{\Theta_{\text{vib}}/T}-1)}$	0	$Nk\ln g_0$
C_V	$\frac{3}{2}Nk$	Nk	$\frac{Nk\Theta_{\text{vib}}^2e^{\Theta_{\text{vib}}/T}}{T^2(e^{\Theta_{\text{vib}}/T}-1)^2}$	0	0
P	$\frac{NkT}{V}$	0	0	0	0

$H=E+pV=E+NkT$; $\quad G=A+pV=A+NkT=\sum_i N_i\mu_i$; $\quad C_P=C_V+Nk$; $\quad \Theta_{\text{rot}}=\hbar^2/2Ik$; $\quad \Theta_{\text{vib}}=\hbar\omega/k=h\nu/k$.

molecule by adding ϵ_{corr} to the sum of Eqs. 11-2 and 11-23:

$$\epsilon_{\text{corr}} = -x\hbar\omega(n+\tfrac{1}{2})^2 - Dc\hbar j^2(j+1)^2(2\pi)^{-1}$$
$$- \alpha\hbar c(n+\tfrac{1}{2})j(j+1)(2\pi)^{-1}. \qquad (11\text{-}43)$$

The term in the dimensionless quantity x reflects the anharmonicity of the potential. The term in D is from centrifugal stretching. The term in α gives the vibration-rotation interaction. The three parameters x, D, and α can be adjusted independently to fit experimental data, or they can be interrelated through a theoretical analysis.*

INTERNAL ROTATION COMPLICATIONS

As mentioned previously, it is usually a complicated task to determine the $3n-5$ (linear molecule) or $3n-6$ (nonlinear molecule) normal modes of vibration of a polyatomic molecule. It sometimes happens that one of these normal modes corresponds to *rotation* of two parts of the molecule with respect to each other about the single bond joining the parts. The classic example is ethane, whose two methyl groups can rotate about the bond H_3C–CH_3. Because organic chemists knew that rotation about this bond was rapid compared with macroscopic times of measurement, statistical mechanicians originally replaced the q_{vib} of the form of Eq. 11-28 from this mode by a classical rotational contribution,

$$q_{\text{int rot}} = \frac{2I_r kT}{\sigma\hbar^2}. \qquad (11\text{-}44)$$

*Detailed corrections to thermodynamic quantities due to rotation-vibration interaction, first- and second-order anharmonicities, and doubly degenerate vibrations are given by H. W. Woolley, *J. Res. Nat. Bur. Stand.* (U.S.), **56**, 105 (1956). The effect of centrifugal distortion on G, S, and C_V is treated by E. B. Wilson, Jr., *J. Chem. Phys.*, **6**, 526 (1936). He obtains corrections of $-\rho RT^2$, $2\rho RT$, and $2\rho RT$ respectively, where ρ is a constant (deg^{-1}) characterizing each molecule. Examples follow: for H_2O, $\rho = 2.04\times10^{-5}$; for NH_3, $\rho = 1.45\times10^{-5}$; for C_2H_4, $\rho = 0.79\times10^{-5}$. The correction to C_V for water at 100 C is about 0.5%. Interesting tables of the corrections to thermodynamic properties of CO_2 due to anharmonicity and vibration-rotation interaction are given by H. W. Woolley, *J. Res. Nat. Bur. Stand.* (U. S.), **52**, 289 (1954). He also shows how the naturally occurring mixture of isotopes has different properties from $^{12}C^{16}O_2$.

Here I_r is the *reduced moment of inertia* characterizing the internal rotation,* and σ is the symmetry number for the internal rotation. In the case of ethane, σ is 3. Symmetry numbers are discussed in the next section of this chapter.

As it became possible to compare accurate experimental results with the theory, discrepancies appeared. The entropy calculated using Eq. 11-44, and even the $\frac{1}{2}Nk$ contribution to C_V, proved to be incorrect. It was reasoned that perhaps internal rotation was hindered by repulsion of the rotating groups. In ethane, hydrogens on one carbon atom will repel their opposites on the other carbon atom most strongly when they are directly opposed (eclipsed), the least when they are staggered. The potential energy is expected to vary with relative angle of rotation ϕ according to a curve shaped something like Fig. 11-6. This could be represented analytically by either of the two rather simple, one-parameter potentials,

$$U(\phi) = \tfrac{1}{2}U_0(1 - \cos 3\phi) \qquad \text{or} \qquad U(\phi) = U_0 \sin^2(\tfrac{3}{2}\phi). \qquad (11\text{-}45)$$

The barrier height U_0 cannot be obtained from molecular quantum mechanics because its calculation is too difficult. Thus it is from experiment, usually with Raman or infrared spectroscopy, that values of U_0 are obtained.† Sometimes U_0 is simply fitted as a last, catchall, adjustable parameter used to make calorimetric results agree with statistical mechanical theory. As such, it has been remarkably successful.

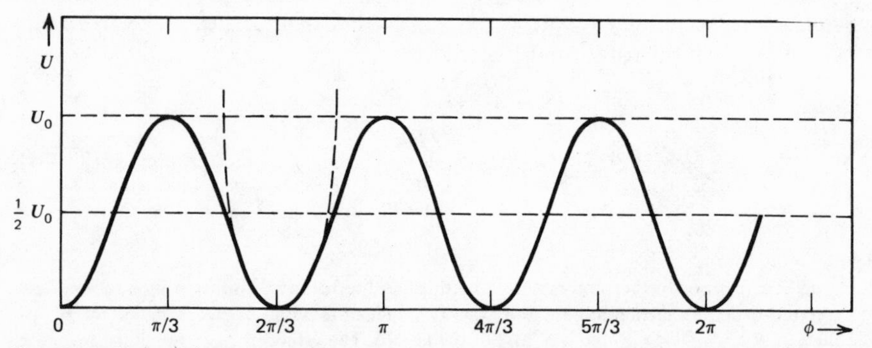

Fig. 11-6. Potential energy as function of angle of rotation about the C–C single bond in ethane, CH_3-CH_3.

*Discussion and references are given by Janz, p. 26.

†E. B. Wilson, Jr. in *Advances in Chemical Physics*, Vol. 2, I. Prigogine, ed. (Interscience, New York, 1959), p. 367.

At temperatures so low that $kT \ll U_0$, the potential of Fig. 11-6 may be approximated closely by the parabola indicated by the dashed line. In that case, the internal "rotation" is simply another normal mode of vibration. Only when the temperature becomes high enough to permit rotation does one need to pay separate attention to internal rotation. Also, at high temperatures, $kT \gg U_0$, the parts barely feel the hindering potential, and rotation about the bond is essentially free. Then one can use Eq. 11-44 with a change in zero-point energy, which is the average height in Fig. 11-6. Problems arise for intermediate temperatures.* The energy levels well below U_0 resemble vibration, those well above U_0 resemble rotation. For molecules with internal rotation one has cause for concern if the overall moment of inertia of the molecule changes in the course of the internal rotation. For example, 1–2 dichloroethane, $H_2ClC{-}CH_2Cl$, has different overall moments of inertia when the two chlorines are next to each other and when they are opposed.

NUCLEAR SPIN COMPLICATIONS AND SYMMETRY NUMBERS

We have ignored one property of atomic nuclei—namely, that they have a magnetic dipole moment due to a nuclear spin angular momentum, characterized by the integral or half-integral quantum number i. Each nucleus has its nuclear spin, determined by which isotope of which element it is. It is impossible to change this spin without a profound nuclear reaction. A given value of i leads to a nuclear degeneracy of $2i+1$; all the $2i+1$ states for the nucleus have almost identical energy. Thus if we were to treat nuclear spins, every single atomic state previously counted in the expression for q would have to be viewed as $2i+1$ different states. Thus the q for any molecule computed previously would be multiplied by a new factor $q_{\text{nuc spin}}$, which would include a term $(2i+1)$ for each atom in the molecule. This would modify the entropy and free energies of every system by an amount that depended solely on what atoms its molecules contained. For any *process*, then, which did not involve transmutation, the *changes* in these quantities would not be affected by the inclusion of $q_{\text{nuc spin}}$. Nuclear spin states essentially just change the zero of entropy; they may be eliminated during the process of choosing a zero for entropies, as discussed in connection with Eq. 6-19. One is thus justified in neglecting the direct effect of nuclear spins on the value of q.

One feature in which nuclear spin is involved is the quantum mechanics leading to the rotational symmetry number σ, introduced earlier. From the

*See Janz, p. 30. It is worth working out in detail the examples presented there.

classical viewpoint it is fairly clear why σ arises as it does. Consider a molecule in which all *like* atoms are numbered and distinguishable. The **symmetry number** σ is defined to be the number of positions that this molecule may be rotated into that differ from each other by only the numbers labeling the like atoms. The classical rotational partition function in the absence of σ counts the whole 4π steradians (sr) of solid angle in which the molecule may rotate. Yet since quantum mechanics demands that like particles be truly indistinguishable, only $1/\sigma$ of the 4π sr represent physically distinguishable alignments. Each position of the molecule within any part of that $4\pi/\sigma$ sr would be identical to and thus indistinguishable from σ other similar positions within the total possible solid angle of 4π sr. Thus only $1/\sigma$ of the classical value of q_{rot} can actually represent the classical limit of the proper quantum mechanical q_{rot}. Classical rotation is discussed in Chap. 18. In this regard, the introduction of $1/\sigma$ into q_{rot} resembles the introduction of $1/N!$ into the expression $Q = q^N/N!$

If the same problem is approached on a purely quantum mechanical basis, certain of the rotational states that would otherwise appear are ruled out for reasons of symmetry. The symmetry arises from a combination of factors, including the nuclear spin. The question of just which rotational levels are allowed and which are not is a delicate one,* and it is important at low temperatures, comparable to Θ_{rot}. The peculiar properties of ortho- and parahydrogen arise from this cause.†

PROBLEMS

11-1. What are the degeneracies of the following rotational energy levels: (*a*) 0, (*b*) $\hbar^2/I$, (*c*) $6\hbar^2/I$, (*d*) $20k\Theta_{rot}$?

11-2. What is the ratio of populations of N_2 molecules vibrationally unexcited to those in the first excited vibrational state at 25 C?

11-3. What fraction of diatomic molecules have rotational energy greater than kT?

11-4. What is the ratio of populations of N_2 molecules rotationally unexcited to those with energy equal to that of the first excited rotational level at 25 C?

11-5. Verify Eq. 11-2 with help of the references given.

11-6. Verify Eq. 11-23 with help of the references given.

11-7. Verify the classical mechanical results for the simple harmonic oscillation of a diatomic molecule, summarized in Eq. 11-17, with help of the references given.

*Mayer and Mayer, pp. 135–138 and 172–178; Fowler and Guggenheim, p. 84.

†G. S. Rushbrooke, *Introduction to Statistical Mechanics* (Oxford University Press, London, 1949), pp. 100–109.

11-8. Calculate the exact expression for E_{rot} in terms of T and Θ_{rot} at temperatures so low that only the ground state and the three lowest excited energy levels are significantly populated.

11-9. Prove that Eqs. 11-14 follow from Eq. 11-13.

11-10. Verify that as the temperature is increased, E_{vib} approaches the line NkT and not the line $NkT+\frac{1}{2}Nk\Theta_{vib}$.

11-11. Using the data given in the paragraph just following Eq. 11-42, find the electronic contributions to E, A, μ, S, C_v, and P for NO as a function of T.

11-12. Sketch a plot of the molar heat capacity of carbon monoxide gas as a function of temperature over the temperature range 0 to about 4000 K.

11-13. Calculate q_{rot} for the CN, OH, and ClO radicals at 300 C, given that the bond lengths are 1.157, 0.971, and 1.49 Å, respectively.

11-14. Verify the entries in the rotation column of Table 11-2.

11-15. Verify the entries in the vibration column of Table 11-2.

11-16. Verify the entries in the zero of energy column of Table 11-2.

11-17. Verify the entries in the electronic column of Table 11-2.

11-18. What is the molar entropy of helium gas at 298.16 K and 1 atm pressure?

11-19. Discuss and compare the results you would expect for the thermodynamic functions E, H, A, G, μ, S, C_V, and C_P for a mole of the following ideal gases: (*a*) Ne at 400 K, (*b*) Ne at 4000 K, (*c*) HBr at 400 K, (*d*) HBr at 4000 K. The degeneracy of the ground state is unity. Utilize Table 11-2. Do not calculate these quantities in detail. What approximation might break down?

11-20. An ideal gas mixture in a volume V is composed of N_a molecules of type a and N_b molecules of type b. Formulate the partition function.

11-21. Find ΔS for the process of combining N_a molecules of type a in a volume V_a with N_b molecules of type b in a volume V_b to form a final mixture of volume V_a+V_b. Do the same for the case in which the molecules of type a and type b are identical.

11-22. Consider N molecules adsorbed on a surface of area $\mathcal{Q}$. They may move freely in the x and y directions but have no freedom in the z direction. Calculate the translational partition function for the system. The system must perform work to increase the surface area by $d\mathcal{Q}$. The part of this work due to the adsorbed molecules is $-\mathcal{P}\,d\mathcal{Q}$. Find the equation of state, which relates $\mathcal{P}$, $\mathcal{Q}$, and T for noninteracting molecules.

11-23. We claim that q_{elect} at low temperatures is g_0, but is it? Consider as simple a case as a gas of hydrogen atoms. The electronic states of a free hydrogen atom are known, their energies being given by $\epsilon_n = -a/n^2$. The quantum number n may have the values 1,2,3,.... If there are two possible spin alignments, g_0 for this atom is 2. What is the value of q_{elect} at any finite temperature using these energy levels? Discuss the meaning of this. Since the probability that the atom will be in a given state is $e^{-\beta\epsilon_{state}}/q_{elect}$, with the value of q_{elect} we have found, what happens to the probabilities? Even of the ground state? What has gone wrong?

11-24. At room temperatures, what contributions to C_V from translation, rotation, vibration, and free rotation about the single bonds would you expect for cadmium dimethyl vapor, CH_3–Cd–CH_3?

11-25. What are the relative populations of the first three rotational energy levels of HCl at 300 K? What are the relative populations of the first three vibrational levels of HCl at 300 K? At what temperature are 10% of the HCl molecules excited vibrationally?

11-26. During the absorption of infrared radiation by room temperature HCl, the molecules are initially in an equilibrium distribution of quantum states. What does that imply about the values of j and of n for the HCl molecules? Absorption of a photon increases n by 1 and simultaneously either increases or decreases j by 1. What set of energies are absorbed by HCl? The probability of a transition is proportional to the population of the state being excited. Predict the intensities of the various lines in the HCl infrared spectrum, and compare your prediction with the actual spectrum, pictured in G. M. Barrow, *Physical Chemistry*, p. 264.

11-27. How many normal modes of vibration has the water molecule? Sketch them and indicate which, if any, you would expect to be degenerate. Do the same for the carbon dioxide molecule and for the acetylene molecule.

11-28. Work out Example 2.3 of Janz in detail, calculating the hindered internal rotation contribution to the entropy of C_2H_5CN at different temperatures.

11-29. Work out Example 2.5 of Janz in detail, calculating the entropy of mixing of both optical and rotational isomers of methylhydrazine.

11-30. A reference says the vibrational spacings of Br_2 and H_2 are 323 and 4395 cm^{-1}, respectively, H_2 being the largest for any known molecule. What is meant by these numbers? Are they consistent with the data in Table 11-1? At 25 C what is the ratio of the number of Br_2 molecules with $n=1$ to the number with $n=0$? What is the comparable ratio for H_2?

12 CHEMICAL EQUILIBRIUM

CRITERION FOR CHEMICAL EQUILIBRIUM

Equilibrium constants for chemical reactions are sometimes obtained more accurately using statistical mechanics than by measurement. In a reaction, one set of molecules is formed as another set is used up; this process is rigorously governed by the stoichiometry, represented by the balanced chemical reaction. For example, during the reaction between H_2 and O_2 to form H_2O, the change in the number of H_2 molecules is always the negative of the change in the number of H_2O molecules. The change in number of O_2 molecules is always half the change in number of H_2 molecules. We might choose to write the reaction as $2H_2+O_2=2H_2O$, in which case we would say that the **stoichiometric number** for H_2O, ν_{H_2O}, is $+2$; for H_2 it is -2; and for O_2 it is -1. Stoichiometric numbers are conventionally chosen to be positive for products of the reaction as written and negative for the reactants. Thus exactly the same chemical reaction is described by writing $H_2O=H_2+\frac{1}{2}O_2$, but the stoichiometric numbers instead would be $\nu_{H_2O}=-1$, $\nu_{H_2}=1$, $\nu_{O_2}=\frac{1}{2}$. For a given balanced chemical equation, the stoichiometric number for compound i is written ν_i. In the following discussion we take the reaction to be written so that there are no fractional values of ν_i.

For a particular chemical reaction, there is an initial set of numbers N_i^0 of molecules of the various reactant and product species present in the volume. Consider the partition function

$$Q=\sum_{\substack{\text{all}\\\text{states}}} e^{-\beta E_{\text{state}}} \tag{12-1}$$

for the system at fixed temperature whose composition is initially fixed by the set of numbers N_i^0. In determining the allowed set of quantum states to use in Eq. 12-1, one must include not only those that would show $N_i=N_i^0$ for each of the various species. After all, the set $N_i=N_i^0+\nu_i$ for all i also represents a condition that can be reached from $N_i=N_i^0$ simply by having the reaction proceed from left to right. The sets $N_i=N_i^0+2\nu_i$, $N_i=N_i^0+$

$3\nu_i$, $N_i = N_i^0 + 4\nu_i$, and so on, can all be reached from $N_i = N_i^0$ by the reaction also; thus they too must be included in the set of states contributing to Q. Similarly, provided N_i^0 is not zero for any of the product molecules, the sets $N_i = N_i^0 - \nu_i$, $N_i = N_i^0 - 2\nu_i$, and so on, must also be included. The conclusion is that *the set of quantum states that must be used in Eq. 12-1 for Q will include every set of molecule numbers N_i that could be generated by the reaction out of the initial molecule numbers N_i^0.*

As an example, suppose the system was initially made up of 100 H_2O molecules, 200 H_2 molecules, and 300 O_2 molecules. The various sets of N_i that must be included in Q are shown in Table 12-1. It would not matter to the ensemble, or to the values of Q or $\ln Q$, which of the columns in Table 12-1 represented the initial state; 100–200–300 generates the same ensemble that would be generated by 102–198–299 or 90–210–305. The column that actually gives the values of the N_i at equilibrium is unknown; it is precisely what we want to know. We do know that whichever column it is, the set of N_i found must be obtainable from the initial 100–200–300 mixture by successive reactions from left to right or from right to left, according to the stoichiometry. The probability of any one column at equilibrium will be faithfully given by the fraction of members of the ensemble that would reflect that set of N_i. Statistical mechanics will not be able to say precisely which column represents the system. The system will be fluctuating from one column to the next. However, the ensemble average of the N_i will prove to be an extraordinarily reliable prediction of the numbers actually found in the system at equilibrium.

Table 12-1. Various Sets of N_i Included in Q for a Mixture Initially Containing $N^0_{H_2O} = 100$, $N^0_{H_2} = 200$, $N^0_{O_2} = 300$

N_{H_2O}	...	90	92	94	96	98	100	102	104	106	108	110	112	114 ...
N_{H_2}	...	210	208	206	204	202	200	198	196	194	192	190	188	186 ...
N_{O_2}	...	305	304	303	302	301	300	299	298	297	296	295	294	293 ...

Thus if we compute the value of $\ln Q$ for one set of N_i and compute $\ln Q$ for another set that differs from the first by ν_i, we obtain the same result. The mathematical expression of this is important. The first equality of Eq. 8-25 shows that the change in $\ln Q$ caused by changing the number of molecules of type i by ν_i is $\nu_i(\partial \ln Q / \partial N_i)$. Thus the change in $\ln Q$ caused by changing the number of each type of molecule by its stoichiometric number must be the sum over all i of $\nu_i(\partial \ln Q / \partial N_i)$. This sum must, then, equal zero:

$$\sum_i \nu_i \left(\frac{\partial \ln Q}{\partial N_i} \right) = 0. \tag{12-2}$$

If this equation is multiplied through by $-kT$, the result is

$$\sum_i \nu_i \left[-kT \left(\frac{\partial \ln Q}{\partial N_i} \right) \right] = 0, \tag{12-3}$$

$$\boxed{\sum_i \nu_i \mu_i = 0.} \tag{12-4}$$

The identification of the chemical potential was made using Eq. 8-20. This is the basic equation of chemical equilibrium, the criterion for equilibrium in a chemically reacting system under the conditions of a canonical ensemble. It has followed from the very basic considerations of this chapter and is true in any system in which the different chemical species can be told apart, regardless of whether the system is solid, liquid, or gaseous. We have as yet made no restrictions on the validity of our results.*

ACTIVITIES AND THE EQUILIBRIUM CONSTANT

In the equilibrium mixture, the chemical potentials are generally functions of the concentrations of the various species, of the environment of the molecules, and of all kinds of other things. Mathematically, one may put all this complexity into a single defined quantity, the *activity of substance i*. This is done by first choosing arbitrarily, but for convenience, a reference condition or **standard state** for substance i whose chemical potential if i were in that condition would be μ_i^0. The μ_i^0 is of course given by nature as soon as the standard state is specified,† and μ_i in the chemically reacting mixture of interest is also specified by nature. The **activity** a_i of substance i is no more than the function of composition, temperature, and other

*The parallel formulation and discussion of Eq. 12-4 in thermodynamics is given by Andrews, pp. 225–228.

†For choices of standarad state, see Andrews, pp. 230–231; or G. N. Lewis, M. Randall, K. S. Pitzer, and L. Brewer, *Thermodynamics*, 2nd ed. (McGraw-Hill Book Co., New York, 1961), Chap. 20.

properties, that satisfies the defining equation

$$\mu_i = \mu_i^0 + kT \ln a_i. \tag{12-5}$$

The advantage in working with a_i instead of μ_i is that a_i often depends quite simply on such measurable properties as concentration, partial pressure, or density, whereas μ_i usually depends on these quantities in a much more complicated way.

Use of Eq. 12-5 in Eq. 12-4 immediately gives an expression for the **equilibrium constant**:

$$\sum_i \nu_i \mu_i^0 + kT \sum_i \nu_i \ln a_i = 0, \tag{12-6}$$

$$\sum_i \nu_i \mu_i^0 = -kT \ln(a_1^{\nu_1} a_2^{\nu_2} a_3^{\nu_3} \cdots), \tag{12-7}$$

$$\boxed{\sum_i \nu_i \mu_i^0 \equiv \Delta G^0 = -kT \ln K.} \tag{12-8}$$

In Eq. 12-8 we simply defined $\Sigma_i \nu_i \mu_i^0$ to be ΔG^0, recognizing that μ_i^0 is the molecular Gibbs free energy of substance i in its standard state. Thus $\Sigma_i \nu_i \mu_i^0 = \Delta G^0$ is the change in Gibbs free energy for a process that uses up the stoichiometric numbers of molecules of the reactants from vessels in which they are pure and in their standard states, forming the stoichiometric numbers of molecules of the products in vessels in which they are pure and in their standard states. In Eq. 12-8 we also identified the product of terms $a_i^{\nu_i}$ in Eq. 12-7 to be the familiar equilibrium constant, expressed in terms of activities rather than concentrations, partial pressures, or other familiar quantities. It is called a constant because its value, $e^{(-\Delta G^0/kT)}$ depends on temperature and choice of standard states only, and not in the least on how much of the various species was initially placed in the reaction vessel. An example might clarify the meaning of the abstract symbolism of Eqs. 12-7 and 12-8: given the reaction $2H_2O = 2H_2 + O_2$, we have $\nu_{H_2} = 2$, $\nu_{O_2} = 1$, and $\nu_{H_2O} = -2$. The value of ΔG^0 is $2\mu_{H_2}^0 + \mu_{O_2}^0 - 2\mu_{H_2O}^0$. This value equals $-kT \ln K$, where K is the equilibrium constant expressed in activities (i.e., $K = a_{H_2}^2 a_{O_2} a_{H_2O}^{-2}$). At equilibrium, this product of activities raised to their stoichiometric numbered power always has the value given by $e^{(-\Delta G^0/kT)}$.

The equilibrium constant K is obtained directly from the chemical potentials of the pure reactants and pure products in their standard states. The energies of the various quantum states for the reactant and product molecules, which are needed to find the standard state partition functions

Q_i^0, thus μ_i^0, often can be found directly from spectroscopic studies and a knowledge of quantum mechanics. Sometimes one finds μ_i^0 not by differentiating Q_i^0 with respect to N_i but by separately finding h_i^0 and s_i^0, the partial molecular enthalpy and entropy respectively, of i in its standard state. Then μ_i^0 is found as $h_i^0 - Ts_i^0$:

$$\Delta G^0 \equiv \sum_i \nu_i \mu_i^0 \equiv \sum_i \nu_i (h_i^0 - Ts_i^0) \equiv \Delta H^0 - T\Delta S^0. \tag{12-9}$$

Since the standard states can be defined at will, they are usually chosen to permit the ready identification of the μ^0's. If a gas is involved, its standard state is usually set as 1 atm pressure and hypothetical *ideal* condition. Whether the gas no longer exhibits ideal behavior at 1 atm pressure (because of interactions between its molecules) is immaterial. One can *imagine* its ideal behavior by neglecting intermolecular interactions and use that as the defined standard state. For such a gas μ_i^0 is easily calculated from Q_i^0, as discussed in Chaps. 10 and 11.

Once K has been found, a problem remains, however. The equilibrium constant is given in terms of activities, which are not directly measurable but are only defined by Eq. 12-5. The results of physical measurements are values of some property like pressure, density, mole fraction, or molar concentration. Thus the problem is to relate the activities to one of these experimentally meaningful quantities: in this matter, statistical mechanics becomes very difficult when the final equilibrium condition is neither an ideal mixture of gases nor an ideal solution. Often, the relationship between experimental quantities and activities must just be obtained empirically. The ideal gas mixture, discussed next, is most important as a starting point for more difficult situations.

EQUILIBRIUM IN AN IDEAL GAS MIXTURE

In an ideal mixture of ideal gases, the gas molecules exist in the confining volume completely unaware that other species are present. Different species do not interact with one another in any way. Thus the chemical potential of species i is obtained from T, V, and N_i, regardless of the numbers of other species present, and Eq. 9-30 is valid for each species separately:

$$\boxed{\mu_i = kT \ln \frac{N_i}{q_i},} \qquad \text{(ideal gas mixture)} \tag{12-10}$$

where q_i is the particle partition function for molecules of type i. Equation 12-10 holds for either a pure ideal gas or for the ith component of an ideal mixture of ideal gases.

Choosing a standard state for an ideal gas contained in a volume V at temperature T is equivalent to choosing a number N_i^0 of molecules of type i which are in the volume when the standard condition is realized. Generally, N_i^0 is chosen by demanding 1 atm pressure in the standard state:

$$1\text{ atm} = \frac{N_i^0 kT}{V} \qquad \text{or} \qquad N_i^0 = \frac{(1\text{ atm})V}{kT}, \tag{12-11}$$

in which case N_i^0 is the same for all species. Since μ_i^0 is given by Eq. 12-10 with N_i^0 inserted for N_i, we can express K in terms of statistical mechanics using Eq. 12-8:

$$\sum_i \nu_i \mu_i^0 = kT \sum_i \nu_i \ln \frac{N_i^0}{q_i} = -kT \ln K, \tag{12-12}$$

$$\ln K = -\sum_i \ln\left(\frac{N_i^0}{q_i}\right)^{\nu_i}, \tag{12-13}$$

$$K = \left(\frac{q_1}{N_1^0}\right)^{\nu_1}\left(\frac{q_2}{N_2^0}\right)^{\nu_2}\left(\frac{q_3}{N_3^0}\right)^{\nu_3}\cdots. \qquad \text{(any gas reaction)} \tag{12-14}$$

This expresses K for a gas reaction in terms of the particle partition functions of the reactants and products. The result, Eq. 12-14, is valid regardless of whether the actual gas mixture is ideal. The problem is that K is a constant only when activities are used, as shown by Eqs. 12-7 and 12-8, and the activity therefore must be expressed in terms of a measurable quantity if K is to prove useful.

For an ideal gas mixture, the relation between a_i and easily measured quantities is immediate. We simply cast Eq. 12-10 into the form of Eq. 12-5 and see what that gives for a_i:

$$\mu_i = kT \ln \frac{N_i}{q_i} = kT \ln \frac{N_i^0 N_i}{N_i^0 q_i}, \tag{12-15}$$

$$\mu_i = kT \ln \frac{N_i^0}{q_i} + kT \ln \frac{N_i}{N_i^0}. \qquad \text{(ideal gas mixture)} \tag{12-16}$$

Comparison of Eq. 12-16 with Eq. 12-5 shows an exact term-by-term correspondence, with a_i given by

$$a_i = \frac{N_i}{N_i^0}. \qquad \text{(ideal gas mixture)} \qquad (12\text{-}17)$$

When this value is combined with the conventional choice of N_i^0, Eq, 12-11, we have

$$a_i = \frac{N_i kT}{(1 \text{ atm}) V} = \frac{P_i}{1 \text{ atm}}, \qquad (12\text{-}18)$$

which illustrates the well-known fact that *the activity of a species in an ideal gas mixture is its partial pressure measured in atmospheres.*

Perhaps the most intuitive way of writing and remembering the results of this section is to start with the definition of the equilibrium constant

$$K \equiv a_i^{\nu_1} a_2^{\nu_2} a_3^{\nu_3} \cdots, \qquad (12\text{-}19)$$

and to substitute from Eq. 12-14 on the left-hand side and from Eq. 12-17 on the right:

$$\left(\frac{q_1}{N_1^0}\right)^{\nu_1}\left(\frac{q_2}{N_2^0}\right)^{\nu_2}\left(\frac{q_3}{N_3^0}\right)^{\nu_3} \cdots = \left(\frac{N_1}{N_1^0}\right)^{\nu_1}\left(\frac{N_2}{N_2^0}\right)^{\nu_2}\left(\frac{N_3}{N_3^0}\right)^{\nu_3} \cdots. \qquad (12\text{-}20)$$

Since all the N_i^0 cancel out, for equilibria in ideal gas mixtures we have

$$\boxed{N_1^{\nu_1} N_2^{\nu_2} N_3^{\nu_3} \cdots = q_1^{\nu_1} q_2^{\nu_2} q_3^{\nu_3} \cdots.} \qquad \text{(ideal gas reactions)} \qquad (12\text{-}21)$$

This result is independent of choice of N_i^0. Let us define K_N to be the left-hand side of Eq. 12-21, an obvious expression of the form of an equilibrium constant except in terms of particle numbers. Similarly, let us define K_q to be the right-hand side of Eq. 12-21. Then the form of Eq. 12-21 is easily remembered:

$$\boxed{K_N = K_q.} \qquad \text{(ideal gas reactions)} \qquad (12\text{-}22)$$

Starting with this, other useful equilibrium constants can easily be obtained. For example, K_ρ, where the density $\rho_i = N_i / V$, is found from Eq.

12-21 by substituting $\rho_i V$ for N_i:

$$(\rho_1 V)^{\nu_1}(\rho_2 V)^{\nu_2}(\rho_3 V)^{\nu_3}\cdots = K_q, \tag{12-23}$$

$$K_\rho = K_q V^{-\Delta\nu}, \qquad \text{(ideal gas reactions)} \tag{12-24}$$

where we have written

$$\sum_i \nu_i \equiv \Delta\nu \tag{12-25}$$

as the change in the total number of moles for the reaction as written. Similarly, K_P can be found by substituting $P_i V/kT$ into Eq. 12-21 for N_i:

$$\left(\frac{P_1 V}{kT}\right)^{\nu_1}\left(\frac{P_2 V}{kT}\right)^{\nu_2}\left(\frac{P_3 V}{kT}\right)^{\nu_3}\cdots = K_q, \tag{12-26}$$

$$K_P = K_q\left(\frac{kT}{V}\right)^{\Delta\nu}. \qquad \text{(ideal gas reactions)} \tag{12-27}$$

DISCUSSION

The simplicity of Eq. 12-21 or 12-22 is striking. In the idealized chemical reaction $r \rightleftharpoons p$, the numbers of reactant and product molecules at equilibrium are governed by

$$\frac{N_p}{N_r} = \frac{\displaystyle\sum_{\text{states of } p} e^{-\beta\epsilon_{\text{state}}}}{\displaystyle\sum_{\text{states of } r} e^{-\beta\epsilon_{\text{state}}}}. \tag{12-28}$$

The reaction will favor products or reactants depending on the relative values of q for product and reactant molecules. One of the species may have a larger q for either of two reasons: it may have *lower lying energy levels,* making individual terms $e^{-\epsilon/kT}$ large. It also may have a *great many states* whose energies may not be so low that each $e^{-\epsilon/kT}$ term is huge but compensate because so many terms contribute to the sum over states. Low-lying states are especially important at low temperatures where the energy difference between reactant and product molecules is large compared with kT. At these low temperatures, the effect of a high density of states is damped by the exponential. But as the temperature is increased, the exponential distribution becomes more uniform, and species with a high density of states begin to predominate. The latter effect is one of increased entropy—loss of specific information about the state the system is in as a great many states become about equally probable. Thus it is not the energy (or enthalpy or "heat content") that determines the equilibrium state, but a balance between that property and entropy. Equation 12-8 reveals that this balance involves the relative values of $\Delta G^0 = \Delta H^0 - T\Delta S^0$; thus it is the quantity $H - TS$ that governs the reaction equilibrium.

A good analogy to the equilibrium between reactants and products is a room filled with hopping frogs.* Suppose the floor is painted part red and part purple. Each unit area of the red region represents a state of the reactants, each unit of the purple area represents a state of the products. The relative populations of molecules present as reactants or products is given by the relative populations of frogs in the two regions. The hopping of the frogs is analogous to the random Brownian-type motion of molecules; jumping from one unit area to another represents transition of a molecule from one state to another; jumping from a region of one color into a region of another represents the formation of products out of reactants, or vice versa. So long as the elevations of the two regions are the same, the equilibrium density of frogs will be the same in both regions and the equilibrium population of frogs in each region will simply be proportional to the area of the region. Thus at equilibrium we have

$$\frac{N_p}{N_r} = \frac{p\text{ area}}{r\text{ area}} = \frac{\text{number of states that represent } p}{\text{number of states that represent } r}. \tag{12-29}$$

This result is identical to Eq. 12-28 when all states of both r and p have the same energy. It shows the effect of entropy on the equilibrium distribution of molecules between reactants and products.

Next, suppose the entire purple region was depressed below the red by a distance h such that energy $\epsilon = mgh$ was required for a frog to hop up from p to r. The equilibrium density of frogs will now be lower in the high-energy red region and higher in the purple. This is because the frogs are jumping with a distribution of energies (or heights) over a range from small to large. Thus only a fraction of the jumps a given frog might make could conceivably take him up from purple to red; the rest of his jumps will not be high enough. However, any jump is enough to get a frog off the edge of the red region and down into the purple. The greater the energy difference between the two states, the more favored is the one with lower energy. Entropy still plays its role, since the number of frogs in each region is the

*Andrews, pp. 94–95.

frog density times the area of the region. But the densities differ in the two regions because of their different energies.

This analogy is useful even quantitatively. Let the density of frogs (frogs per unit area) in the purple and red regions be given by ρ_p and ρ_r, respectively. Let ν be the probability that a given frog near the boundary between r and p jumps toward the opposite-colored region in unit time. This probability increases with temperature as the frogs get more lively, but it does not depend on whether the frog is initially in r or p, since each frog likes both regions equally well and thus behaves the same in both. Let $P(r \to p)$ be the probability that a properly directed jump from r toward p will in fact land the frog in p. Since we have r higher than p, $P(r \to p) = 1$. Similarly, let $P(p \to r)$ be the probability that a properly directed jump from p toward r will land the frog in r. This is less than 1 because not all jumps can be high enough. Suppose $P(p \to r)$ is the exponential $e^{-(\epsilon_r - \epsilon_p)/kT}$.

Now, the number of frogs in the area A near the boundary jumping from r toward p in unit time is the number of such frogs ρ_r^A times the jumping probability ν. The number of such jumps that successfully put the frog into p is this times $P(r \to p)$. Similarly, the rate of successful jumps from p into r is given by $\rho_p A \nu P(p \to r)$. At equilibrium, the rate of jumps from r into p is just balanced by the rate from p into r, and we have

$$\rho_r A \nu P(r \to p) = \rho_p A \nu P(p \to r) \tag{12-30}$$

$$\frac{\rho_p}{\rho_r} = \frac{1}{P(p \to r)} = e^{(\epsilon_r - \epsilon_p)/kT}. \tag{12-31}$$

We now can find the ratio of N_p to N_r:

$$\frac{N_p}{N_r} = \frac{\rho_p A_p}{\rho_r A_r}, \tag{12-32}$$

where A_p is the area of the purple region and A_r the area of the red. These areas are just the number of quantum states that represent p and r, respectively; thus by using Eq. 12-31 we obtain

$$\frac{N_p}{N_r} = \frac{A_p}{P(p \to r) A_r} = \frac{(\text{number of } p \text{ states}) e^{-\epsilon_p/kT}}{(\text{number of } r \text{ states}) e^{-\epsilon_r/kT}}. \tag{12-33}$$

This result is precisely what is given by Eq. 12-28 for the case of all p states having the energy ϵ_p and all r states the energy ϵ_r.

Consider the role played by temperature. As the temperature is increased, the frogs become more active. They not only jump more often, which increases the *rate* of the process, but higher jumps become more favored. Great heights, which are almost impossible at low temperatures, become common, and the little jumps characteristic of low temperatures become relatively rare. Thus as T increases, the frogs become less aware of the energy difference between the states, because their average jump is more than enough to clear the barrier. Therefore, at high temperatures the effect of entropy predominates; the two regions might just as well have been at the same height. At low temperatures, however, once the frogs fall down into the region with lower energy, they have great trouble clearing the barrier, no matter how high it is. When T is small, therefore, the effect of energy predominates.

One last manipulation of Eq. 12-33 is interesting: Equation 6-26 may be used to replace A_p by $e^{(S_p/k)}$ and A_r by $e^{(S_r/k)}$:

$$\frac{N_p}{N_r} = \frac{e^{S_p/k - \epsilon_p/kT}}{e^{S_r/k - \epsilon_r/kT}} = e^{(T\Delta S^0 - \Delta\epsilon^0)/kT}, \qquad (12\text{-}34)$$

which is clearly the same as Eq. 12-8 for an ideal gas reaction for which $\Delta\nu = 0$ (hence $\Delta(p^0V) = 0$, and $\Delta A^0 = \Delta G^0$).

The same analogy is suitable for illustrating the molecular theory of chemical kinetics. There we are interested in how fast equilibrium is established, given an initial state of disequilibrium, and we must recognize that the boundary between the two regions is not simply a step, as considered previously. Instead, there exists an *activation energy barrier* or wall right at the boundary over which the frogs must jump if they are to land in the opposite region. Whichever way the frogs are jumping, the barrier must be cleared. This requirement changes the value of $P(r\rightarrow p)$ from 1 to $e^{-\epsilon_a/kT}$, where ϵ_a is the height of the activation energy barrier. It also changes $P(p\rightarrow r)$ from $e^{-(\epsilon_r - \epsilon_p)/kT}$ to $e^{-(\epsilon_a + \epsilon_r - \epsilon_p)/kT}$. The net rate from r to p will then be the difference between the left- and right-hand sides of Eq. 12-30, with these new values of the P's. Clearly, the equilibrium result when the net rate vanishes is unchanged by the introduction of the activation energy barrier. The study of the kinetics is more difficult than the study of the equilibrium, since kinetics requires knowing the values of A and of ν. Clearly, ν is a rapidly increasing function of the temperature.

It is sometimes asked why certain processes driven by ΔG go to completion and others do not. For example, a liquid in a closed container evaporates until the pressure reaches its vapor pressure, at which time equilibrium is reached. Reactants mixed either at constant pressure or in a fixed volume undergo chemical reaction until product concentrations become large enough for equilibrium to be reached. On the other hand, a liquid above its boiling point at fixed pressure vaporizes completely. The difference is that in the first two processes, as the reaction proceeds, the activity of products builds up while the activity of reactants stays almost the same (vapor pressure) or decreases (chemical reaction). In the frog analogy, as frogs cross from r to p, the frog concentration in p increases, and the rate of the reverse process therefore increases until finally the forward and reverse rates become equal. In a phase transition at constant pressure, however, the activities of reactants and products remain unchanged as the reaction proceeds; only the relative amounts vary. As frogs hop from r to p, A_r is decreased and A_p increased in such a way that $\rho = N/A$ for both reactants and products stay constant. Thus equilibrium cannot be reached, and the phase transition proceeds until the supply of frogs in r is exhausted and $A_r = 0$.

PRACTICAL CALCULATIONS OF EQUILIBRIUM CONSTANTS

Following are the usual sources of information needed to calculate the q's in the expression for K for a gas reaction. For q_{trans} the molecular weight and volume are trivial. For q_{rot} for either overall or internal rotation, the moment(s) of inertia and symmetry number(s) are found from a molecular model, usually based on electron and x-ray diffraction studies, although sometimes spectroscopic evidence is also used. These experiments furnish the necessary bond lengths and bond angles to construct the model. For q_{vib} the set of fundamental vibration frequencies (with anharmonicity and centrifugal corrections and vibration-rotation interaction) comes from spectroscopy, usually infrared and Raman. For q_{elect} the degeneracy of the ground state (and energies and degeneracies of any other electronic states with energies low enough to contribute appreciably) comes from ultraviolet and visible spectroscopy. Most of the above-mentioned information is of extraordinary precision, because spectroscopic measurements are some of science's most precise. However, for q_{zero} the dissociation energy D_0 is obtained either from spectroscopy or from calorimetry, and it is not easy to obtain with any degree of precision.*

Because of the difference in precision between the value of D_0 and the rest of the data, it is customary to tabulate separately the contributions to thermodynamic variables that come from the precise data and those from D_0. This assures that the inaccuracies in D_0 values do not contaminate the more precisely known contributions. Thus instead of tabulating the Gibbs

*The many problems involved in determining D_0 are discussed by A. G. Gaydon, *Dissociation Energies*, 2nd ed. (Chapman and Hall, London, 1953).

free energy G_T^0 and the enthalpy H_T^0 of pure compounds at the temperature T, one tabulates the *difference* between G_T^0 or H_T^0 and the quantity $H_{T^0}^0$. Here, $H_{T^0}^0$ is the enthalpy of formation of the pure compound from its elements at a reference temperature T^0, usually chosen to be either 0 or 298.15 K. For the choice of $T^0 = 0$ K, these differences simply involve the subtraction of $-D_0$ from G_T^0 or H_T^0. This is clear from Table 11-2. For ideal gases with the choice of $T^0 = 298.15°$ K, the quantity subtracted is more complicated, but it still predominantly includes D_0. The choice of standard state implied by the superscript 0 is the hypothetical ideal gas at 1 atm pressure.

It is convenient for variables being tabulated as function of temperature to be slowly varying, thus permitting easy interpolation with a minimum of entries. Since division of the above-mentioned differences by T makes them vary more slowly with temperature, tabulations are commonly made of the so-called **enthalpy function**

$$\frac{H_T^0 - H_{T^0}^0}{T} \tag{12-35}$$

and the **free energy function**

$$-\left(\frac{G_T^0 - H_{T^0}^0}{T}\right) = S_T^0 - \left(\frac{H_T^0 - H_{T^0}^0}{T}\right). \tag{12-36}$$

Of course to find G_T^0 or H_T^0, the best available values of $H_{T^0}^0$ must be tabulated, too. In addition, tabulations are usually made of a specific heat and of the entropy, neither of which involves D_0.*

*Once one knows how to calculate the thermodynamic functions from statistical mechanics, tabulations of them can be employed with confidence and understanding. Extensive tabulations based on the most precise spectroscopic and calorimetric data available have been made. Most common inorganic compounds will be found in the compilation edited by D. R. Stull, *JANAF Thermochemical Tables,* 2nd ed. (Washington, D.C., U.S. Bureau of Standards, 1972). This monumental work is a promising start toward a complete tabulation of the thermodynamic properties of all known compounds, which is a major goal of physical chemistry. For organic compounds one may consult D. R. Stull, E. F. Westrum, Jr., and G. C. Sinke, *The Chemical Thermodynamics of Organic Compounds*, (New York, John Wiley & Sons, 1969). These books contain references to other collections, and they have useful discussions of the problems inherent in using tabulated thermodynamic data. Many tables of thermodynamic data of various kinds, expecially handy for quick reference, are given by G. N. Lewis, M. Randall, K. S. Pitzer, and L. Brewer, *Thermodynamics*, 2nd ed., (New York, McGraw-Hill Book Co., 1961); see Subject Index under "Tables."

PROBLEMS

12-1. Express the equilibrium constant for particle densities for the ideal gas reaction $AB \rightleftharpoons A + B$ in terms of the atomic masses m_A and m_B, the characteristic temperatures for the AB molecule Θ_{rot} and Θ_{vib}, the ground state degeneracies $g_{0,A}$, $g_{0,B}$, and $g_{0,AB}$, and D_0 for the AB molecule. Be sure to use the same zero of energy for q_{AB} as for q_A and q_B. At temperatures well below Θ_{vib}, how does the equilibrium constant change with increasing T? For *very* high temperatures, what temperature dependence does your equation give? If this seems to be nonsensical, what assumption made in its derivation has broken down?

12-2. Contrast the result of Prob. 12-1 with what would be found for the reaction $A_2 \rightleftharpoons 2A$.

12-3. Express the equilibrium constant for particle densities for the ionization of a monatomic gas, $A \rightleftharpoons A^+ + e^-$. Assume no potential energy of interaction between the separated A^+ and e^-. The degeneracies of the three species are $g_{0,A}$, g_{0,A^+}, and 2, respectively. Consider only the ground electronic state of the A atom, the ionization energy for which is ϵ^0.

12-4. In the ionization treated in Prob. 12-3, define the temperature Θ_{ion} by the equation $\epsilon^0 = k\Theta_{\text{ion}}$. Estimate the value of T/Θ_{ion} at which the equilibrium constant has all three densities identical and with reasonable values, say, $10^{20}/\text{cm}^3$. The g_0's may be set equal to unity. A typical ionization energy is 12 eV, and 1 eV $= 1.6 \times 10^{-12}$ erg. A mole of electrons weighs about $1/1840g$.

12-5. The energy required to ionize the hydrogen atom, $H \rightleftharpoons H^+ + e^-$, is 13.53 eV. The atom is doubly degenerate, but the ion has $g_0 = 1$. If an initial concentration of hydrogen atoms of 0.01 mole/liter is heated to 10,000 K, what fraction of the atoms are ionized? Assume the fraction ionized to be much less than unity.

12-6. Using data from Table 11-1, calculate the equilibrium constant in terms of densities for the ideal gas reaction $H_2 + Cl_2 \rightleftharpoons 2HCl$ at 800C. All species are singly degenerate.

12-7. Work out Example 2.1 in Janz, in detail. This is the calculation of thermodynamic properties from spectroscopic data for trifluoroacetonitrile, CF_3CN, and it is an excellent example of the power of statistical mechanics.

12-8. Work through the paper by E. A. Guggenheim, *Trans. Faraday Soc.*, **37**, 97 (1941), where K_P is calculated at a number of temperatures for the hydrogenation of ethylene to give ethane. The partition function is obtained from spectral data, and the resulting calculated values of K_P are probably better than the experimental values. Some helpful comments on Guggenheim's paper are given by M. Dole, *Introduction to Statistical Thermodynamics* (Prentice-Hall, Englewood Cliffs, N. J., 1954).

12-9. Work through in detail pages 163–170 of Dole (see Prob. 12-8), where the calculation of isotopic exchange reaction equilibrium constants is discussed.

12-10. Work through the paper by G. Waddington and coworkers, *J. Am. Chem. Soc.*, **71**, 797 (1949). This is interesting because it presents the thermodynamic and spectroscopic data for thiophene side by side, allowing comparison of the results.

12-11. In terms of the frog model developed in this chapter, justify the known temperature dependence of the solubility of liquids and solids in liquids as contrasted with that of the solubility of gases in liquids.

12-12. Three-fourths of naturally occurring Cl atoms have mass number 35 and the rest have mass number 37. These atoms form Cl_2 molecules almost at random. What distribution of molecular weights will be found in naturally occurring Cl_2 gas? What is the entropy of one mole of this Cl_2 gas at 1 atm and 25 C compared with the pure, separated components of the mixture, all at 1 atm and 25 C? Here is a case of the entropy of a substance depending on whether one knows the substance to be a mixture of isotopes. If balances were so inaccurate that the isotopes of Cl had not been discovered, Cl_2 gas would be assigned a different entropy. How would that affect the results of thermodynamic predictions of the properties of Cl_2?

12-13. Calculate K_P for the dissociation of Na_2 to form Na atoms at 800 K. Spectroscopy gives a dissociation energy of 0.73 eV, a fundamental vibration frequency of 159.23 cm^{-1}, and an internuclear distance of 3.078 Å. The atoms are doubly degenerate, the molecule is nondegenerate.

12-14. Calculate K_P for $H_2+D_2=2HD$ at 27 C. The vibration frequencies of H_2, HD, and D_2 are 4371, 3786, and 3092 cm^{-1}, respectively. The moments of inertia are 0.458, 0.613, and 0.919×10^{-40} $g\text{-}cm^2$.

12-15. Estimate K_P for ${}^{16}O_2+{}^{18}O_2=2{}^{16}O{}^{18}O$ if the three different molecules have nearly the same Θ_{vib}, Θ_{rot}, and mass.

12-16. Using the data in Table 11-1, calculate K_P for $Cl_2=2Cl$ at 10^3 K. The ground state of Cl_2 is singly degenerate, Cl atoms are doubly degenerate.

12-17. A vessel contains oxygen at 3750 K and 1.70 atm total pressure. For the reaction $O_2=2O$, K_P is 0.85 atm at this temperature. What fraction of O_2 molecules are dissociated? What fraction of undissociated O_2's have vibrational energy greater than 0.2 D_0 (see Table 11-1)? Why is dissociation so greatly favored over vibrational excitation?

12-18. What are ortho- and parahydrogen? What is the ratio of ortho to para in H_2 at room temperature, and why? What is this ratio at temperatures below 80 K, and why?

12-19. Suppose there are two isotopes of the same element weighing m_A and m_B. The element forms a diatomic ideal gas. Let z be the fraction of atoms of type A in the natural abundance. At equilibrium at 25 C and 1 atm total pressure, what will be the mole fractions of A_2, B_2, and AB as function of z? Consider the element to be sufficiently heavy that all three molecules have approximately the same Θ_{vib}, Θ_{rot}, and masses. How do these mole fractions depend on temperature? On pressure?

12-20. It has been said that precision in measurement of thermodynamic properties is vital for accurate predictions. If ΔH^0 for a chemical reaction is measured with an error of 2.7 kcal, how much will the calculated value of the equilibrium constant be in error at 25 C?

PART 3

GENERAL QUANTUM STATISTICAL THERMODYNAMICS

13 PERFECT CRYSTALS

GENERAL THEORY

Mechanically, a perfect crystal is very much like a single, extremely large molecule. Suppose the crystal consists of a total of N atoms or monatomic ions. If each of these particles is bound into the crystal lattice, the total internal energy of the crystal must lie in the contributions from the many vibrational modes of the particles. The position of each particle is specified by three coordinates; thus the N particles introduce $3N$ degrees of freedom into the crystal. Three of these are specified by fixing the center of mass of the crystal and three by orienting its axes in space. Associated with these six coordinates are the translational and rotational energy of the crystal as a whole. Such energies are treated by classical mechanics and do not normally hold statistical mechanical interest.

This leaves $3N-6$ coordinates that can be specified independently to make the crystal a completely determined mechanical entity. Choosing these $3N-6$ coordinates such that their motions represent independent simple harmonic vibrations is the major problem in treating crystals. Such coordinates are called the **normal coordinates** of the crystal, analogous to the normal coordinates for the vibrational modes in a molecule. Associated with each normal coordinate is a characteristic angular velocity (sometimes called frequency) ω_i and an energy contribution of the form of the quantum mechanical harmonic oscillator, Eq. 11-23:

$$\epsilon_i = (n_i + \tfrac{1}{2})\hbar\omega_i = (n_i + \tfrac{1}{2})h\nu_i. \tag{13-1}$$

There will be $3N$ of these frequencies ω_i (for macroscopic crystals where $N > 10^{20}$, 6 is utterly negligible compared with $3N$), and N energies, a different one associated with each normal mode. The theory of lattice dynamics is devoted to the problem of determining how the values of these $3N$ characteristic frequencies are distributed.

We shall carry the statistical mechanical problem along as far as possible before worrying about the frequencies. The partition function is the sum over states,

$$Q = \sum_{\text{states}} e^{-\beta E_{\text{state}}}. \tag{13-2}$$

Just as we found in Chap. 11, Q factors into a separate vibrational contribution from each of the $3N$ normal modes plus a separate electronic contribution from each particle. There is no division of the result by $N!$ in the case of crystals because specification of a set of N different particle electronic states gives the states of the particles *located at the N lattice points,* and the lattice points are distinguishable by virtue of their location in the crystal. Thus the question of which particle of a group of like particles is at which lattice point is meaningless, but the lattice points themselves are distinguishable. This matter has already been encountered in the discussion of residual entropy in Chaps. 6 and 7. In addition, if the N particles comprising the crystal were molecules instead of atoms or monatomic ions, there would be a product of contributions to Q for the normal modes of vibration of each of the molecules in the crystal (and for rotation in addition, if the molecules were still rotating in the solid state). Furthermore, if there were free electrons in the crystal, as there are in metals, these would have to be treated separately in a fashion similar to that described in Chap. 14. The treatment of this chapter would then be valid only for the lattice of ions in the metal.

Let us focus on Eq. 13-2 in which the independent normal modes of vibration contribute all the quantum states for the crystal. The energy in each mode is given by Eq. 13-1, and the total energy of the crystal present in Eq. 13-2 is the sum of Eq. 13-1 over all the normal modes: $E_{\text{state}} = (n_1 + \frac{1}{2})\hbar\omega_1 + (n_2 + \frac{1}{2})\hbar\omega_2 + (n_3 + \frac{1}{2})\hbar\omega_3 + \cdots$. Since changing the quantum number n_i for a single mode changes the state of the entire crystal, the partition function factors into a *product* of q_{vib} terms for the various normal modes:

$$Q = \left(\sum_{n_1=0}^{\infty} e^{-\beta(n_1+1/2)\hbar\omega_1} \right) \left(\sum_{n_2=0}^{\infty} e^{-\beta(n_2+1/2)\hbar\omega_2} \right) \cdots, \tag{13-3}$$

$$Q = \left(\frac{e^{-(1/2)\beta\hbar\omega_1}}{1 - e^{-\beta\hbar\omega_1}} \right) \left(\frac{e^{-(1/2)\beta\hbar\omega_2}}{1 - e^{-\beta\hbar\omega_2}} \right) \cdots, \tag{13-4}$$

where we simply referred to the derivation of Eq. 11-28 from Eq. 11-25.

Thus the logarithm of Q is given by the *sum*

$$\ln Q = -\sum_{i=1}^{3N} \frac{\hbar\omega_i}{2kT} - \sum_{j=1}^{3N} \ln(1 - e^{\hbar\omega_j/kT}), \tag{13-5}$$

where we have written a single sum over two terms as two separate sums. From Eq. 13-5 all thermodynamic properties of a perfect crystal could be calculated if only the distribution of the normal frequencies ω_i were known. The bonds holding together a crystal lattice are some 100 to 1000 times weaker than typical chemical bonds that hold molecules together. Since chemical bonds are so strong that $\hbar\omega = kT$ at temperatures of, say, 3000 K (see Table 11-1), the weaker forces in a crystal typically give frequencies about 10 times lower (see Eqs. 11-16 and 11-22). Thus at room temperature and below, the normal modes in typical crystals have $\hbar\omega$ of the same order of magnitude as kT.

EINSTEIN'S THEORY

The simplest approximation to the actual frequency distribution in a crystal is that proposed as an approximate treatment by Einstein in 1907. He took all normal frequencies in a given crystal to be the same:

$$\omega_i = \omega_E, \qquad \text{the same for all } i. \tag{13-6}$$

Einstein's proposal works at temperatures so high that all vibrational modes are excited, because just how they came to be excited—one after another, as the temperature increased—would then be of little importance. The use of an average frequency as representative of all the frequencies would not be too unreasonable. A physical model could be given by imagining each of the $3N$ normal modes to be the vibration of one of the N particles about its equilibrium position in one of the three directions. Each particle should experience a similar average environmental potential energy, approximately the same in all three directions. Then all normal frequencies would be identical. However, this model is a great oversimplification at lower temperatures.

The results of Einstein's theory are certainly easy to obtain: If Eq. 13-6 is used to define a characteristic temperature Θ_E:

$$\hbar\omega_E = k\Theta_E, \tag{13-7}$$

then $\ln Q$ in Eq. 13-5 becomes

$$\ln Q = -\frac{3N\Theta_E}{2T} - 3N\ln(1-e^{-\Theta_E/T}). \tag{13-8}$$

All thermodynamic quantities of interest can be calculated from Eq. 13-8.

In particular, the heat capacity is interesting to compare with experiment:

$$E = kT^2\left(\frac{\partial \ln Q}{\partial T}\right)_{V,N}, \tag{13-9}$$

$$= kT^2\frac{3N\Theta_E}{2T^2} + kT^3 3N\frac{e^{-\Theta_E/T}\Theta_E}{(1-e^{-\Theta_E/T})T^2}, \tag{13-10}$$

$$E = \frac{3}{2}Nk\Theta_E + \frac{3Nk\Theta_E}{e^{\Theta_E/T}-1}; \tag{13-11}$$

$$C_V = \left(\frac{\partial E}{\partial T}\right)_{V,N} = 3Nk\left(\frac{\Theta_E}{T}\right)^2\frac{e^{\Theta_E/T}}{(e^{\Theta_E/T}-1)^2}. \tag{13-12}$$

The heat capacity of a crystal, according to Einstein's theory, is the same function of Θ_E/T for all crystals. The parameter Θ_E will of course have different values for crystals composed of different substances. Experimentally, it is true that plots of C_V versus T do have similar shapes for monatomic crystals and may often be superimposed by proper scaling of the temperature. At high temperatures, Θ_E/T is small and Eq. 13-12 becomes

$$C_V \to 3Nk, \qquad T\to\text{high compared with } \Theta_E. \tag{13-13}$$

This condition has long been recognized experimentally as the **law of Dulong and Petit.** It is by no means a surprising result, since it is just the classical limit of the contribution to the heat capacity of N simple harmonic oscillators (see Chap. 18). At low temperatures, however, where C_V is sensitive to the fraction of the normal modes that are excited, the limit of Eq. 13-12,

$$C_V \to 3Nk\left(\frac{\Theta_E}{T}\right)^2 e^{-\Theta_E/T}, \qquad T\to\text{low compared with } \Theta_E, \tag{13-14}$$

approaches zero too rapidly for decreasing T, because of the rapidly damping exponential. Experimentally, C_V goes to zero as T^3 at sufficiently low temperatures. It is understandable that a model assigning the same characteristic temperature to all normal modes would have C_V approach zero too fast. It is very difficult to excite any of the modes at temperatures much below Θ_E. Thus C_V would emerge lower than if one correctly assigned a range of characteristic temperatures from the very low to the moderately high. The low value of C_V reflects the fact that very little energy input would raise the temperature considerably, since there are almost no sinks for heat. Thus Einstein's model shows C_V to start out almost zero, to rise rapidly when T increases to Θ_E, then to level off at $3Nk$ for high temperatures. Such behavior is qualitatively in agreement with experiment.

DEBYE'S THEORY

In 1912 Peter Debye proposed a *distribution* of normal frequencies instead of a single frequency for all modes. This led to the proper T^3 dependence of C_V at low temperatures and to the Dulong-Petit value of $3Nk$ at high temperatures. Debye's theory has proved most useful in correlating theory with experiment. In trying to find a set of $3N$ low-frequency normal modes, he presumed that proper combinations of these low-frequency vibrations could reproduce any vibrational motion in the crystal. He proposed that these low-frequency modes were analogous to the vibrations by which *sound* waves are transmitted in a crystal. Thus the problem is to determine the distribution of frequencies of the $3N$ lowest-lying sound waves of velocity C in a crystal of volume V. There are many ways to do this, but it is instructive first to formulate the differential equation for the motion of a sine wave.

Let us define a simple **wave** as any function of position that retains its shape but moves at a constant velocity. At time zero this function is $\Phi_0(x)$, where for simplicity we consider only one dimension. Let this disturbance be moving at velocity C in the plus-x direction. An observer watching at x will see the disturbance moving from left to right. At time t, he sees what was located a distance Ct to his left at time zero. Therefore, the time dependence of the disturbance of an arbitrary wave at point x is given simply by

$$\Phi(x,t)=\Phi_0(x-Ct). \tag{13-15}$$

Now the normal modes of vibration must be simple harmonic motion of the form treated in Eqs. 11-15 to 11-22. Thus $\Phi_0(x)$ will have the form of a sine or cosine:

$$\Phi_0(x) = A\cos\left(\frac{2\pi x}{\lambda}\right). \tag{13-16}$$

It is clear that λ is the **wavelength** of the distrubance, since Eq. 13-16 takes the same value at any two positions that differ from each other by an integral multiple of λ. The space and time dependence of Φ is given by Eqs. 13-15 and 13-16:

$$\Phi(x,t) = A\cos\frac{2\pi(x - Ct)}{\lambda} = A\cos\left(\frac{\omega x}{C} - \omega t\right). \tag{13-17}$$

In the last step we noted that C/λ is the number ν of complete cycles per second, and of course $\omega = 2\pi\nu$. Not all frequencies are permitted for Eq. 13-17 because the wave must have the value zero at the surface of the crystal. The problem of finding the distribution of frequencies ω that give Eq. 13-17 zero value at the walls is perhaps most simply done by analogy with the solution of the Schrödinger wave equation for a particle in a box (Eq. 10-4). The time-independent differential equation satisfied by Eq. 13-17 is found by differentiating that equation twice with respect to x:

$$\frac{\partial\Phi}{\partial x} = -\left(\frac{\omega}{C}\right)A\sin\left(\frac{\omega x}{C} - \omega t\right); \qquad \frac{\partial^2\Phi}{\partial x^2} = -\left(\frac{\omega}{C}\right)^2 A\cos\left(\frac{\omega x}{C} - \omega t\right). \tag{13-18}$$

Identifying Φ itself as the cosine term yields

$$\frac{\partial^2\Phi}{\partial x^2} = -\frac{\omega^2}{C^2}\Phi \quad \text{or} \quad \nabla^2\Phi = -\frac{\omega^2}{C^2}\Phi, \tag{13-19}$$

where the second writing is the three-dimensional analog of the first. Equation 13-19 is often called the **Helmholtz equation**; we want to solve it subject to the boundary condition $\Phi = 0$ at the walls. This is precisely the mathematical problem posed by Eq. 10-4:

$$\nabla^2\psi = -\frac{2m\epsilon}{\hbar^2}\psi. \tag{13-20}$$

Since these two problems, with boundary conditions, are identical, we can replace the $2m\epsilon/\hbar^2$ of the Schrödinger equation with the ω^2/C^2 of the

Helmholtz equation in the solution, Eq. 10-5:

$$\epsilon(n_x,n_y,n_z)=\frac{\pi^2\hbar^2}{2ma^2}(n_x^{\,2}+n_y^{\,2}+n_z^{\,2}), \tag{13-21}$$

$$\frac{2m\epsilon}{\hbar^2}=\frac{\pi^2}{a^2}(n_x^{\,2}+n_y^{\,2}+n_z^{\,2}), \tag{13-22}$$

$$\frac{\omega^2}{C^2}=\frac{\pi^2}{a^2}(n_x^{\,2}+n_y^{\,2}+n_z^{\,2}) \tag{13-23}$$

Thus there is a frequency ω for every choice of the positive integers n_x, n_y, and n_z. The result is simpler in the quantum number space of Fig. 10-1:

$$\frac{\omega^2}{C^2}=\frac{\pi^2}{a^2}n^2; \qquad \frac{\omega}{C}=\frac{\pi}{a}n; \qquad \frac{d\omega}{C}=\frac{\pi}{a}\,dn. \tag{13-24}$$

The number of quantum states (thus the number of ω's) within a spherical shell lying between n and $n+dn$ is the volume of the positive octant of such a shell (see the discussion of Eqs. 10-8 and 10-9):

$$\frac{1}{8}\,4\pi n^2\,dn=\frac{1}{8}\,4\pi\left(\frac{\omega^2a^2}{C^2\pi^2}\right)\left(\frac{a\,d\omega}{\pi C}\right)=\frac{V\omega^2\,d\omega}{2\pi^2C^3}\,. \tag{13-25}$$

Thus Eq. 13-25 gives the number of normal modes of vibration with frequencies between ω and $\omega+d\omega$ if sinusoidal waves propagate within a volume V at speed C. For lattice vibrations, there are three possible polarizations: two transverse waves at right angles to each other and one longitudinal (this contrasts with sound waves, which are longitudinal only). If all three polarizations have the same speed C (or if their average speed is taken to be C), all that remains is

$$\text{Number of modes with frequency between } \omega \text{ and } \omega+d\omega=\frac{3V\omega^2\,d\omega}{2\pi^2C^2}, \tag{13-26}$$

for the frequency distribution. This is the basic result of the Debye theory.

Once Eq. 13-26 is adopted for the frequency distribution, the partition function, Eq. 13-5, generates the values of the various thermodynamic properties. The only thing still needed is to *cut off the total number of normal modes at 3N*. Debye chose the $3N$ different normal modes, as given by Eq. 13-26, which had *longest wavelengths* or shortest frequencies. Thus all frequencies with ω greater than an upper limit $\omega_{\max}$ are eliminated. The value of ω_{max} is found by equating the number of modes with $\omega < \omega_{max}$ to $3N$:

$$\int_0^{\omega_{max}} \frac{3V\omega^2\, d\omega}{2\pi^2 C^3} = 3N. \tag{13-27}$$

Performing the integral yields the frequency ω_{max} at the Debye cutoff; this is called the **Debye frequency**:

$$\frac{V\omega_{max}^3}{2\pi^2 C^3} = 3N; \qquad \omega_{max} = C\left(\frac{6\pi^2 N}{V}\right)^{1/3}. \tag{13-28}$$

The cutoff frequency can be used to define the **Debye characteristic temperature** Θ_D,

$$k\Theta_D = \hbar\omega_{max}, \tag{13-29}$$

in terms of which some of the formal results appear simpler.

Figure 13-1 plots the frequency distribution of normal modes in a crystal for both the Einstein and Debye theories. The area under either curve between any two values of ω is the number of vibrational quantum states in the crystal that are said to exist by the appropriate theory, whose frequency lies between the two values. The total area under each curve is of course $3N$. A lot of theoretical work has gone into finding more nearly exact distributions for realistic crystal lattice dynamics, but the results have been complicated and somewhat disappointing.*

The value of Θ_D for a crystal,

$$\Theta_D = \frac{\hbar C}{k}\left(\frac{6\pi N}{V}\right)^{1/3}, \tag{13-30}$$

is proportional to the average speed of propagation of vibrations in the crystal and to the cube root of the particle density. Both the density and the hardness decrease as one goes from top to bottom of the periodic table. Thus one would expect Θ_D to get smaller. Values of Θ_D, given in Table 13-1 in the form of a periodic table, confirm these trends. Because the theory is only approximate, different values of Θ_D permit better fits with experimental heat capacity curves over different temperature ranges. This explains why experimenters quote different values of Θ_D for the same

*For references and a brief discussion, see Hill, Sec. 5-4. A general reference to the entire subject of heat capacities of solids is M. Blackman, *Handbuch der Physik* (Springer, Berlin) **7.1**, 325 (1953).

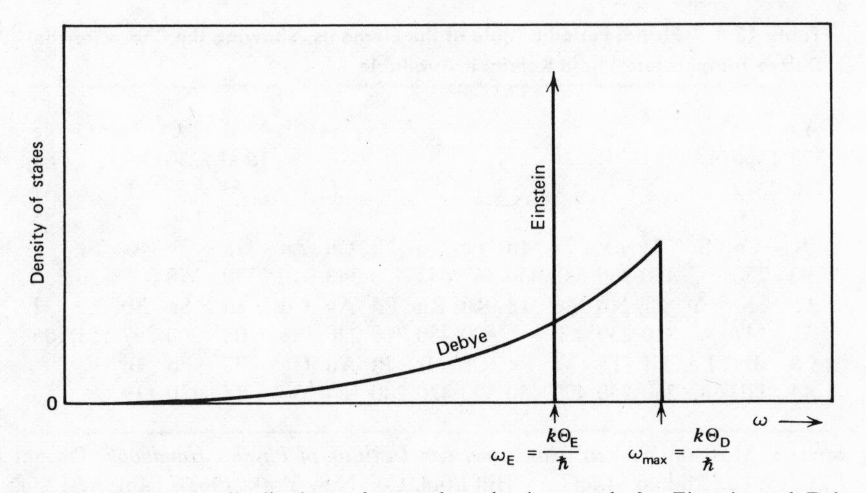

Fig. 13-1. Frequency distributions of normal modes in crystals for Einstein and Debye theories.

crystal. For values of Θ_D small compared with T, the vibrational modes have already thawed out to give the classical Dulong-Petit behavior. At temperatures comparable to or less than Θ_D, the value of C_V will be less than $3Nk$.

The summations over states of Eq. 13-5 are replaced by integrations over ω after multiplication by the density of states as given by Eq. 13-26:

$$\ln Q = -\int_0^{\omega_{\max}} \left(\frac{\hbar\omega}{2kT}\right)\left(\frac{3V\omega^2\,d\omega}{2\pi^2C^3}\right) - \int_0^{\omega_{\max}} \ln(1-e^{-\hbar\omega/kT})\left(\frac{3V\omega^2\,d\omega}{2\pi^2C^3}\right). \tag{13-31}$$

The first term can be integrated directly; the variable change $z=\hbar\omega/kT$ is customarily made in the last term. We write $\ln Q$ in two equivalent forms, which are verified in Probs. 13-4 and 13-5:

$$\ln Q = -\frac{3V\hbar\omega_{\max}^4}{16\pi^2kTC^3} - \frac{3Vk^3T^3}{2\pi^2C^3\hbar^3}\int_0^{\Theta_D/T} \ln(1-e^{-z})z^2\,dz; \tag{13-32}$$

$$\ln Q = -\frac{9N\Theta_D}{8T} - 3N\ln(1-e^{-\Theta_D/T}) + \frac{3NT^3}{\Theta_D^3}\int_0^{\Theta_D/T} \frac{z^3\,dz}{e^z-1}. \tag{13-33}$$

Table 13-1. Partial Periodic Table of the Elements, Showing the Characteristic Debye Temperature Θ_D in Kelvin, if Available

Li	Be											B	C	N	O	Ne
370	1160											1000	2230			61
Na	Mg											Al	Si	P	S	Ar
158	400											428	640			85
K	Ca	Sc	Ti	V	Cr	Mn	Fe	Co	Ni	Cu	Zn	Ga	Ge	As	Se	
90	230		420	360	630	450	467	445	450	343	310	320	370		90	
Rb	Sr	Y	Zr	Nb	Mo	Tc	Ru	Rh	Pd	Ag	Cd	In	Sn	Sb	Te	I
52	147		310	230	450		600	480	300	226	188	108	200	207	153	106
Cs	Ba	La	Hf	Ta	W	Re	Os	Ir	Pt	Au	Hg	Tl	Pb	Bi		
44	110	142	260	240	400	430	500	420	240	164	80	87	110	119		

Source: Most data taken from *American Institute of Physics Handbook*, Dwight E. Grey, ed., 2nd ed, (McGraw Hill Book Co., New York, 1963). Table 4e-12, p. 4–61. Carbon is given for the diamond form. Graphite has value 420.

In Eq. 13-33, the integral of Eq. 13-32 is performed by parts, using Eq. B-11, which is verified in Prob. 13-5.

All thermodynamic properties of an ideal crystal can be calculated as functions of Θ_D from Eq. 13-32 or 13-33. We summarize some of those properties here, leaving verification of these results to the problems:*

$$E=\frac{9}{8}Nk\Theta_D+\frac{9NkT^4}{\Theta_D^3}\int_0^{\Theta_D/T}\frac{z^3\,dz}{e^z-1}, \tag{13-34}$$

$$C_V=\frac{36NkT^3}{\Theta_D^3}\int_0^{\Theta_D/T}\frac{z^3\,dz}{e^z-1}-\frac{9Nk\Theta_D}{T(e^{\Theta_D/T}-1)}, \tag{13-35}$$

*The definite integrals in Eqs. 13-32 to 13-37 can be related to one another but cannot be simplified further. However, since each of them has been tabulated as function of (Θ_D/T), the work is finished when any calculation has been reduced to something involving one of those integrals. Using tables of values of such definite integrals is in no way inferior to using tables of values of $\ln x$ ($\ln x$ being defined as the value of $\int_1^x z^{-1}dz$). The integrals are usually called the **Debye function** $D(\Theta_D/T)$, but different authors call different integrals by that name, and one must always be careful to establish which definition is used by a given author. The two commonest choices are

$$D_E\left(\frac{\Theta_D}{T}\right)\equiv\frac{3T^3}{\Theta_D^3}\int_0^{\Theta_D/T}\frac{z^3\,dz}{e^z-1}; \qquad D_C\left(\frac{\Theta_D}{T}\right)\equiv\frac{3T^3}{\Theta_D^3}\int_0^{\Theta_D/T}\frac{z^4e^z\,dz}{(e^z-1)^2}, \tag{13-38}$$

where we have used subscripts to distinguish between the *Debye energy function* D_E and the *Debye heat capacity function* D_C. Useful tables of the various Debye functions are given by Lewis, Randall, Pitzer, and Brewer (pp. 659–664).

$$C_V = \frac{9NkT^3}{\Theta_D^3} \int_0^{\Theta_D/T} \frac{z^4 e^z\, dz}{(e^z - 1)^2}, \tag{13-36}$$

$$S = \frac{12NkT^3}{\Theta_D^3} \int_0^{\Theta_D/T} \frac{z^3\, dz}{e^z - 1} - 3Nk\ln(1 - e^{-\Theta_D/T}). \tag{13-37}$$

In Fig. 13-2, heat capacity as calculated from the Debye theory is plotted versus temperature. It has both the Dulong-Petit limiting value for high temperatures and T^3 dependence for low temperatures, as the reader will verify in the problems. Both in the T^3 region and in the interpolation region, the Debye theory agrees well with experiment for many crystals.

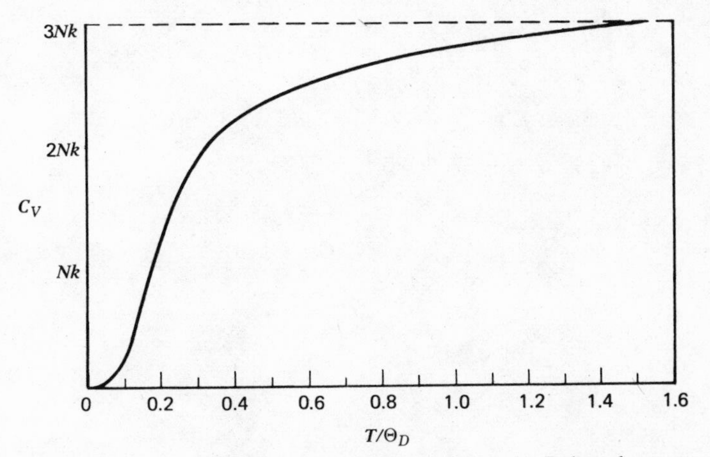

Fig. 13-2. Plot of C_V versus T, as calculated from the Debye theory.

PROBLEMS

13-1. Verify that the low-temperature limit of the Debye heat capacity is indeed the T^3 behavior, $C_V = 12\pi^4 NkT^3/5\Theta_D^3$.

13-2. Verify that the high-temperature limit of the Debye heat capacity is indeed the Dulong-Petit value.

13-3. Calculate the coefficient of thermal expansion $\alpha = V^{-1}(\partial V/\partial T)_P$ for a Debye solid. What is its high-temperature limit?

13-4. Verify Eq. 13-32.

13-5. Verify Eq. 13-33.

13-6. Verify Eq. 13-34.

13-7. Verify Eq. 13-35.

13-8. Verify Eq. 13-36.

13-9. Verify Eq. 13-37.

13-10. Prove that the time dependence of *any* wave, Eq. 13-15, is given by the so-called **wave equation**:

$$\frac{\partial^2 \Phi}{\partial x^2} = \frac{1}{C^2} \frac{\partial^2 \Phi}{\partial t^2}. \tag{13-38}$$

What assumption about the *time* dependence of Φ is necessary to leave the spatial part of Φ a solution of the one-dimensional Helmholtz equation, Eq. 13-19?

13-11. Estimate the specific heat of diamond, silicon, tin, and potassium, using Table 13-1 and Fig. 13-2. The atomic weights are 12, 28, 118.7, and 39.1, respectively. Compare with handbook values.

13-12. Find the equation relating the sublimation pressure for an Einstein solid with the temperature, if the vapor is monatomic and ideal. Take the enthalpy of sublimation to be $\Delta H_s = N \Delta h_s$.

14 METALS: THE ELECTRON GAS

The statistical mechanics of the ideal Fermi-Dirac gas can yield considerable insight into the properties of metals through the use of a greatly oversimplified model of the metallic structure. The metal is treated as if it were composed of the outer valence electrons of its atoms and of the resulting positive ions. The positive ions form a lattice through which the electrons are completely free to move. The plus and minus charges of the ions and electrons are assumed to neutralize each other with only negligible resulting forces on the electrons. Such a model is certainly crude, but as a first approximation to the mechanical structure of metals, its simplicity and predictive successes make it well worth studying.

As mentioned in Chap. 9, electrons are fermions. Whether the electron gas can be described by Boltzmann statistics depends on the value of the ratio r of the number of available particle states to the number of particles. This ratio is correctly given by Eq. 10-10 or 10-11 for electrons, except that since electrons may have spin quantum number either $+\frac{1}{2}$ or $-\frac{1}{2}$, r for electrons is just twice that given by Eq. 10-10. We use that result to define the so-called **characteristic temperature of degeneration** Θ at which r equals unity for an electron gas:

$$\frac{2\pi}{6N}\left(\frac{2Mk\Theta}{N_0}\right)^{3/2}\frac{V}{\pi^3\hbar^3}=1, \tag{14-1}$$

$$\Theta=\frac{3^{2/3}\pi^{4/3}\hbar^2 N_0\rho^{2/3}}{2Mk}. \tag{14-2}$$

For metals, a reasonable estimate of ρ is obtained by assuming that each metal atom furnishes one valence electron and that a molar volume is perhaps 6 cm^3. There would then be about 10^{23} valence electrons per cubic centimeter of the metal. The moleculaı weight of an electron is only about

1/1840; therefore, we have

$$\Theta = \frac{3^{2/3}\pi^{4/3}(1.054\times10^{-27}\,\text{erg-sec})^2(6.02\times10^{23}\,\text{mole}^{-1})(10^{23}\,\text{cm}^{-3})^{2/3}}{2(1840)^{-1}\,\text{g-mole}^{-1}(1.38\times10^{-16}\,\text{erg-K}^{-1})},$$

$$\Theta = 91{,}900\,\text{K} \quad \text{or} \quad \text{about } 10^5\,\text{K}. \tag{14-3}$$

Thus a metal would have to be heated to 10^6 degrees before Boltzmann statistics would yield a good approximation to the partition function of the electron gas.

Room temperatures are so much lower than Θ that they can be treated as minor perturbations from absolute zero. At 0 K the electrons will crowd into the N lowest-lying particle states available. This is clear from the expression for the occupation number

$$N_i = \frac{1}{e^{(\epsilon_i - \mu)/kT} + 1}. \tag{14-4}$$

If ϵ_i exceeds the chemical potential μ_0 at 0 K, the exponential in the denominator is infinitely large and N_i is zero. On the other hand, if ϵ_i is less than μ_0, the exponential vanishes at 0 K and N_i is unity. This case is called **complete degeneration** of the electron gas. All particle states with energy below μ_0 are occupied. It is common to call μ_0 the **Fermi limiting energy,** or the **Fermi level**. Electrons with energy μ_0 are said to lie on the surface of the **Fermi sea**. Electrons with energy less than μ_0 are in the Fermi sea, below its surface by a depth $\mu_0 - \epsilon_i$. In Fig. 14-1, the occupation number is plotted against energy of the states for an electron gas at absolute zero.

The value of μ_0 is the energy of the Nth from the lowest-lying particle state. In the quantum number space of Fig. 10-1, the number of states represented by any volume is just twice the volume, owing to the two

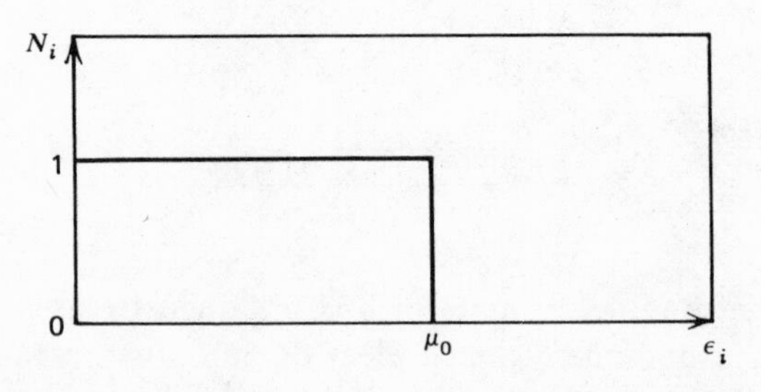

Fig. 14-1. Plot of occupation number against energy of the states for an electron gas at absolute zero.

possible spins. We want to equate twice the volume of the positive octant of a sphere of radius $n_{\max}$ to the value N, to find $n_{\max}$:

$$2 \cdot \tfrac{1}{8} \cdot \tfrac{4}{3}\pi n_{\max}^3 = N, \tag{14-5}$$

$$n_{\max} = \left(\frac{3N}{\pi}\right)^{1/3}. \tag{14-6}$$

Therefore, using Eq. 10-5, we write

$$\mu_0 = \epsilon_{\max} = \frac{\pi^2\hbar^2}{2mV^{2/3}} n_{\max}^2 \tag{14-7}$$

$$\mu_0 = \frac{\pi^2\hbar^2}{2m}\left(\frac{3N}{\pi V}\right)^{2/3} = k\Theta, \tag{14-8}$$

where the identification with Θ was made by comparison with Eq. 14-2.

At 0 K the average energy of an electron will be less than μ_0 because most electrons lie below the surface of the Fermi sea. Here we prove that $\bar{\epsilon}$ is $\frac{3}{5}\mu_0$. The occupation numbers used in the expression for the average energy

$$E = \sum_i \epsilon_i N_i \tag{14-9}$$

are unity for $n < n_{\max}$ and zero for $n > n_{\max}$. Thus Eq. 14-9 is simply the sum of ϵ_i over all states within the eighth of the sphere of radius $n_{\max}$. The energy of such a state is

$$\epsilon(n) = \frac{\pi^2\hbar^2}{2mV^{2/3}} n^2. \tag{14-10}$$

Since there are two possible spins, we have

$$E = 2\int d\mathbf{n} \frac{\pi^2\hbar^2}{2mV^{2/3}} n^2, \tag{14-11}$$

where the notation $d\mathbf{n}$ designates a volume element (e.g., $dn_x\, dn_y\, dn_z$ or $n^2 \sin\theta\, dn\, d\theta\, d\chi$) in quantum number space. The integral is simplest if performed in spherical coordinates. For these, the limits on n are 0 to $n_{\max}$ and the limits on both θ and χ are 0 to $\pi/2$, as is clear from consideration of Fig. 10-1:

$$E = \frac{\pi^2 \hbar^2}{mV^{2/3}} \int_0^{n_{\max}} n^4 \, dn \int_0^{\pi/2} \sin\theta \, d\theta \int_0^{\pi/2} d\chi, \tag{14-12}$$

$$E = \frac{\pi^2 \hbar^2}{mV^{2/3}} \left(\frac{n_{\max}^5}{5} \right) (1) \left(\frac{\pi}{2} \right), \tag{14-13}$$

$$E = \frac{3N\pi^2 \hbar^2}{10m} \left(\frac{3N}{\pi V} \right)^{2/3} = \frac{3}{5} N\mu_0. \tag{14-14}$$

This leads to an enormous value of the pressure, as calculated from Eq. 9-24:

$$P = \frac{2E}{3V} = \frac{2N\mu_0}{5V} = \frac{2}{5}\rho\mu_0 = \frac{2}{5}\rho k\Theta, \tag{14-15}$$

where substitution was made from Eqs. 14-14 and 14-8. Thus the pressure can be estimated in the same way Eq. 14-3 was obtained:

$$P = \frac{2}{5}\rho k\Theta = \frac{2}{5}(10^{23}\,\text{cm}^{-1})(1.38 \times 10^{-16}\,\text{erg-K}^{-1})(10^5 K)$$
$$\times \left(\frac{10^{-6}\,\text{atm}}{\text{dyne-cm}^{-2}} \right) \approx 5 \times 10^5\,\text{atm}. \tag{14-16}$$

This huge pressure is balanced by the attraction between the electrons and the positive ions of the lattice, and the balance holds the metal together.

The foregoing calculations have been made considering the temperature to be zero. Because the ratio T/Θ is so very small, this is an excellent approximation. In particular, the energy E is almost independent of temperature for small T/Θ. Consequently the specific heat of the electrons is negligible—approximately $\frac{1}{2}\pi^2 NkT/\Theta$—because as the temperature is slightly increased, relatively few additional particle states are made available to electrons. Only a few electrons quite near the surface of the Fermi sea can move up into excited states. The bulk of the electrons remain below μ_0 even for temperatures large compared with 300 K but small compared with Θ. Before the advent of quantum mechanics, the failure of the electron gas to contribute to the heat capacity of metals was one of the riddles of statistical mechanics.*

*For example, see J. H. Jeans, *The Dynamical Theory of Gases*, 4th ed., 1925; reprinted by Dover Publications, New York, 1954, Sec. 524.

It is true that at finite temperatures, T/Θ is not truly zero, and the diagram of Fig. 14-1 should look more like that of Fig. 14-2. The width of the range of deviation of the two diagrams is of order kT. Thus the contribution to C_v by the electrons is very small, since so few of them do find enough energy to go into levels above the surface of the Fermi sea.* Only at extremely low temperatures, where the lattice heat capacity (proportional to T^3) has damped out, does the electron heat capacity (proportional to T) become a significant fraction of the total heat capacity.

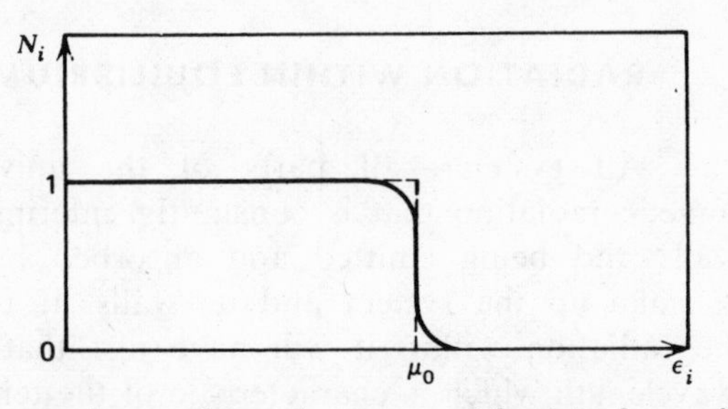

Fig. 14-2. Plot of occupation number against energy of the states for an electron gas at finite temperature $T \ll \Theta$.

PROBLEMS

14-1. Calculate S for the electron gas at 0 K. Is the result surprising?

14-2. Why is Eq. 9-5 consistent with putting $\mu = \epsilon_{max}$ at 0 K?

14-3. At what high temperature does the Debye heat capacity of the sodium ion lattice equal the heat capacity $\frac{1}{2}\pi^2 NkT/\Theta$ from the free electrons? Use the value of Θ obtained in Eq. 14-3 and Θ_D from Table 13-1. Prove that below that temperature the electronic contribution is less than the lattice contribution. Also prove the low-temperature crossing capacities of these heat capacities to be at 0.952 K.

14-4. Each sodium atom (atomic weight 23) contributes one valence electron to metallic sodium (density 0.97 g/cm^3). At what temperature would the number of particle states for electrons equal the number of valence electrons in 1 cm^3 of sodium? How is this result changed for 10 liters of sodium?

14-5. Copper has atomic weight 63.54, density 9 g/ml, melting point 1356.5 K, and one free electron per atom. Find the density of free electrons and the Fermi energy. Is copper degenerate at all temperatures below its melting point?

*This problem is treated in detail by Mayer and Mayer, Chap. 16.

15 RADIATION: THE PHOTON GAS

RADIATION WITHIN EQUILIBRIUM SYSTEMS

All systems—all parts of the universe—are filled with electromagnetic radiation that is constantly entering and departing through the walls and being emitted and absorbed by the atoms or molecules that make up the system and its walls. If the system is in equilibrium, the radiation within it will have just that distribution of intensity and wavelength which is characteristic of the temperature of the system. The particle picture of radiation is of a gas of noninteracting photons, which are bosons of zero rest mass. Unlike material particles, photons may be either created or absorbed by the matter with which they interact. *Any* number of photons *might* be in the system at a given time. When one determines the partition function for the photon gas by summing over all the system's quantum states, one must include *every possible value of* N, from zero to infinity. Only then will Q be a sum over *all* possible states. The immediate result of including terms for all values of N in Q is that there is no longer any N-dependence in Q, since all possible N values are already represented:

$$\mu = -kT\left(\frac{\partial \ln Q}{\partial N}\right)_{T,V} = -\frac{kT}{Q}\left(\frac{\partial Q}{\partial N}\right)_{T,V} = 0. \qquad (15\text{-}1)$$

Thus, μ for a photon gas is zero at equilibrium, a result which follows directly from the fact that photons can be created or destroyed within the system. This result is analogous to the criterion for chemical equilibrium, except that criterion $\sum_i \nu_i \mu_i = 0$ has the stoichiometry that strictly governs the creation and annihilation built in. No such conservation is applicable to the creation or annihilation of photons.

A photon is characterized by its energy $\epsilon = h\nu = \hbar\omega$, where ω is the angular velocity (sometimes called *frequency*) in radians per second. Since μ is zero, the population of photons of a given frequency is given by Eq.

9-13 to be

$$N(\omega)=\frac{1}{e^{\hbar\omega/kT}-1} \tag{15-2}$$

for each possible polarization of a photon. There are two such polarizations; for example, photons could be viewed as representing either right or left circularly polarized radiation, appropriate combinations of which could represent any arbitrary polarization. Or they could be viewed as representing linearly polarized radiation in each of two directions at right angles, appropriate combinations of which could again represent arbitrary polarization.

The number of allowed frequencies is huge; summing over frequencies is always performed by integration. We thus need to determine the density of quantum states, that is, the number of states for photons in a volume V whose frequency lies between ω and $\omega+d\omega$. This requires the relativistic Schrödinger equation, which happens to be precisely the Helmholtz equation, 13-19, whose solution we have already found. Thus the density of states for photons is just twice the value of Eq. 13-25:

$$\text{Number of states with frequency between } \omega \text{ and } \omega+d\omega=\frac{V\omega^2\,d\omega}{\pi^2c^3}. \tag{15-3}$$

Since each state has a population given by Eq. 15-2, the number of photons dN_ω within the frequency range ω to $\omega+d\omega$ is given by the product of Eqs. 15-2 and 3:

$$dN_\omega=\frac{V\omega^2\,d\omega}{\pi^2c^3(e^{\hbar\omega/kT}-1)}. \tag{15-4}$$

The energy dE_ω of the radiation in the frequency range ω to $\omega+d\omega$ is simply the number of photons in that range dN_ω times the energy of each, $\hbar\omega$:

$$dE_\omega=\frac{V\hbar\omega^3\,d\omega}{\pi^2c^3(e^{\hbar\omega/kT}-1)}. \tag{15-5}$$

This result, called **Planck's radiation law**, was very important historically. It was first suggested by Planck in 1900 as "an interpolation formula which resulted in a lucky guess,"* joining the two limits which were at that time well known: $\hbar\omega\gg kT$ (Wien's empirical law) and $\hbar\omega\ll kT$ (the Rayleigh-Jeans law). The interpretation of Planck's radiation law as implying that

*Quoted from Max Planck's Nobel Prize inaugural address.

radiation was emitted by a field of oscillators having *quantized* energies represented the introduction of quantum effects into physics. This opened the door to the rapid development of quantum mechanics and the flowering of modern physics in the twentieth century.

In Fig. 15-1, the change in energy with frequency $dE_\omega/d\omega$ as given by Eq. 15-5, sometimes called the **spectral density**, is plotted as a function of ω. The area under this curve (or integral of Eq. 15-5) between any two values of ω gives the energy of the radiation contained in V at temperature T whose frequency lies between the two values. In Probs. 15-1 and 15-2 it is shown that the maximum value of the spectral density and the frequency

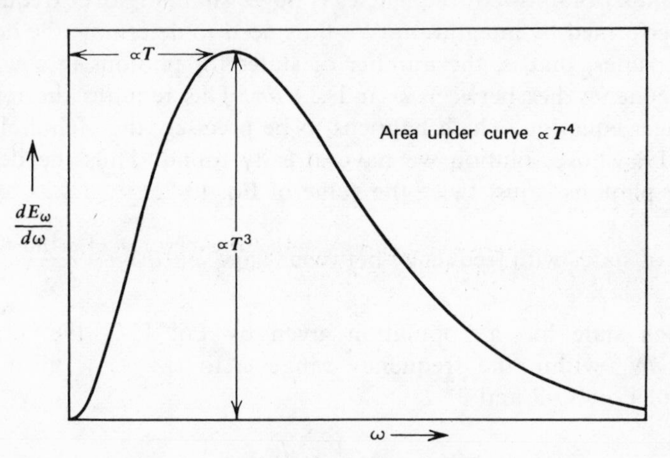

Fig. 15-1. Plot of spectral density against frequency.

$\omega_{\max}$ at which it has that value are given by

$$\left(\frac{\partial E_\omega}{\partial \omega}\right)_{\max} = \frac{1.42\, Vk^3T^3}{\pi^2 c^3 \hbar^2}, \tag{15-6}$$

$$\omega_{\max} = 2\pi\nu_{\max} = 2.822\frac{kT}{\hbar}. \tag{15-7}$$

Equation 15-7 is sometimes called **Wien's displacement law**. As the temperature increases, the bulk of the radiant energy comes from higher frequency radiation (the maximum is displaced toward the blue from the red). The *total* radiant energy E contained in V increases as the fourth power of T:

$$E=\int_0^{\infty}\frac{V\hbar\omega^3\,d\omega}{\pi^2c^3(e^{\hbar\omega/kT}-1)}, \tag{15-8}$$

$$E=\frac{Vk^4T^4}{\pi^2c^3\hbar^3}\int_0^{\infty}\frac{z^3\,dz}{e^z-1}, \tag{15-9}$$

$$E=\frac{V\pi^2k^4T^4}{15c^3\hbar^3}. \tag{15-10}$$

In Eq. 15-9, the variable change $z=\hbar\omega/kT$ was made; in Eq. 15-10, the value of the integral was obtained from Appendix B, Eq. B-9. This result is the so-called **Stefan-Boltzmann law**. In 1884 Boltzmann demonstrated the proportionality of E to T^4 by thermodynamic arguments. It had been shown previously by Stefan on empirical grounds.

The radiation exerts a pressure that can be determined by the method of Prob. 9-4 once the V dependence of the energy levels $\epsilon=\hbar\omega$ is known. This is obtained from Eq. 13-24:

$$\epsilon=\hbar\omega=\frac{\hbar\pi cn}{a}=\hbar\pi cnV^{-1/3}; \tag{15-11}$$

thus the energy levels are proportional to $V^{-1/3}$. The method of Prob. 9-4 then shows

$$PV=\frac{E}{3} \tag{15-12}$$

$$P=\frac{E}{3V}=\frac{\pi^2k^4T^4}{45c^3\hbar^3}=2.4\times10^{-21}T^4 \qquad \text{(in atm).} \tag{15-13}$$

BLACKBODY RADIATION

Suppose one has a system in equilibrium at temperature T contained in thermally insulating walls. A good way to study the radiation inside would be to poke a tiny hole in the wall and observe the radiation emitted. If the hole is small enough, the photon leakage will not appreciably disturb the equilibrium condition within the system. A similar phenomenon for material particles, their *effusion*, is studied in Chap. 18. Any radiation striking the hole from outside will pass in and be absorbed before it has bounced around enough to be reemitted through such a small hole. Since absorbing surfaces are called *black*, the hole is called a **blackbody**, and the radiation effusing out of it is called **blackbody radiation**. We now find the spectral density of this blackbody radiation.

Equilibrium radiation is completely isotropic; that is, photons are equally likely to be going in any direction. Thus the *spectral density per unit volume per unit solid angle* is simply $dE_\omega / d\omega$ divided by $4\pi V$ (4π being the total number of steradians):

$$\frac{1}{4\pi V}\frac{\partial E_\omega}{\partial \omega} = \frac{\hbar\omega^3}{4\pi^3 c^3 (e^{\hbar\omega/kT} - 1)}. \tag{15-14}$$

In the time t, the volume of radiation that leaks out of a hole of area A depends on the angle θ with which the radiation approaches the hole. This is the parallelepiped presented in Fig. 15-2. All the radiation within this parallelepiped with angle from the normal of θ will go through the hole in the time t. The volume of the parallelepiped is $ctA\cos\theta$. If Eq. 15-14 is multiplied by this volume and by the element of solid angle, $\sin\theta\, d\theta\, d\chi$, and divided by t, the result is the spectral density of the radiation in the solid angle denoted by $d\theta\, d\chi$, emitted per second from the hole of area A:

$$\frac{\hbar A \omega^3 \sin\theta \cos\theta\, d\theta\, d\chi}{4\pi^3 c^2 (e^{\hbar\omega/kT} - 1)}. \tag{15-15}$$

The meaning of this quantity is as follows: when multiplied by $d\omega$, Eq. 15-15 is the energy of radiation in the frequency range ω to $\omega + d\omega$ effusing through a hole of area A per second in a direction such that θ lies between θ and $\theta + d\theta$ and χ lies between χ and $\chi + d\chi$. If we are not interested in

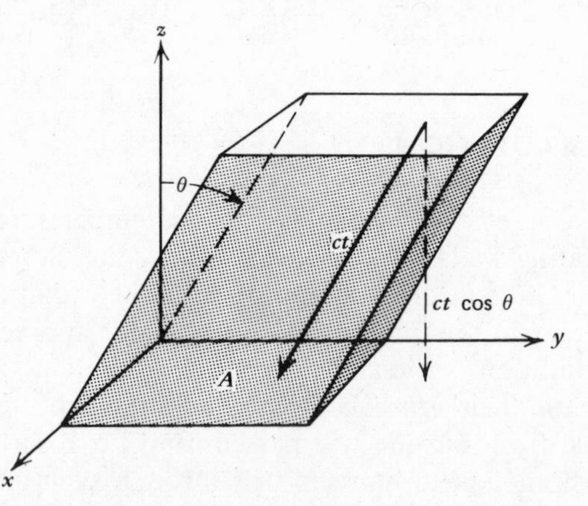

Fig. 15-2. Parallelepiped of impact for photons with a hole of area A.

the angular dependence of the blackbody radiation, we can integrate Eq. 15-15 over angles to get the *spectral density $J(\omega)$ of the total radiation emitted through area A per second:*

$$J(\omega)=\frac{\hbar A\omega^3}{4\pi^3c^2(e^{\hbar\omega/kT}-1)}\int_0^{\pi/2}\sin\theta\cos\theta\,d\theta\int_0^{2\pi}d\chi, \tag{15-16}$$

$$J(\omega)=\frac{\hbar A\omega^3}{4\pi^2c^2(e^{\hbar\omega/kT}-1)}. \tag{15-17}$$

The form of Eq. 15-17 is used experimentally to determine temperatures of furnaces that are too hot to measure otherwise. One measures spectral density as function of frequency for blackbody radiation from the furnace and compares the result with Eq. 15-17. The value of T is determined as that which best fits the experimental curve. The *total blackbody radiation per unit area per second*, J, is $1/A$ times the integral of Eq. 15-17 over all frequencies. By analogy with Eq. 15-8, we can write this immediately:

$$J=\frac{cE}{4V}=\frac{\pi^2k^4T^4}{60c^2\hbar^3}=5.669\,T^4\text{erg}(\text{cm}^{-2})\text{sec}^{-1}\text{K}^{-4}, \tag{15-18}$$

where the last step followed from Eq. 15-10. From an intuitive viewpoint, Eq. 15-18 is interesting. The radiant energy density in the system is E/V. In one second, radiation goes a distance c; thus cE/V is the energy per unit area that would leak out if all of it were going directly toward the hole. The factor $\frac{1}{4}$ accounts for the fact that radiation is going in all directions, not just directly toward the hole. Temperatures of hot objects obtained from Eq. 15-18 by measuring J are called **brightness temperatures**.

PROBLEMS

15-1. Prove Eq. 15-7.

15-2. Prove Eq. 15-6.

15-3. Over which frequency interval is the number of photons greater than the number of states available, making necessary multiple occupancy of states? Over which states is the converse true? Prove that for the latter frequencies, the equivalent of Eq. 15-5 using Boltzmann statistics would be correct.

15-4. What is the radiation pressure at room temperature? What must the temperature be before the radiation pressure becomes 1 atm?

15-5. How hot is a blackbody if its spectral density has a maximum in the yellow at a wavelength of 5890 Å?

15-6. Explain how measurement of the intensity of emission from two different spectral lines in a heated gas could be used to find the temperature of the gas. What further information would be needed? What is the meaning of the "rotational temperature"? The "vibrational temperature"? The "electronic temperature"? Possible references are G. H. Dieke (pp. 9–14) and H. P. Broida (pp. 265–286), in *Temperature—Its Measurement and Control in Science and Industry*. Vol. II. H. C. Wolfe, ed. (Reinhold Publishing Co., New York, 1955).

15-7. Measurement of the temperatures of stars is restricted to studies of the radiation received from these bodies on earth. Resulting values of T are called "color temperatures," "brightness temperatures," "effective temperatures," "excitation temperatures," and other names. What are the meanings of these temperatures? On what assumptions do they depend? How are they related? A possible reference is C. Payne-Gaposchkin, pp. 31–58, in the reference of Prob. 15-6.

15-8. The **emissivity** of a body is the ratio of the radiant energy of wavelength λ (or with wavelength between λ and $\lambda + d\lambda$) emitted by the body to the energy it would emit at that temperature and wavelength if it were a blackbody. Emissivity is a function of both λ and T and is less than 1—for gases, often much less. Do a short literature survey on values of emissivity and of how closely real bodies approach the form of Fig. 15-1. A starting point might be R. J. Thorn and G. H. Winslow in *Temperature—Its Measurement and Control in Science and Industry*. Vol. 3, C. M. Herzfeld, ed. (Reinhold Publishing Co., New York, 1962), pp. 421–447.

16 THE SECOND LAW, FLUCTUATIONS, AND TEMPERATURE

THE SECOND LAW

There are numerous ways to state the second law of thermodynamics, and the statement adopted is almost a matter of taste. The author prefers not to drag in the entire universe (on which very few experiments have been carried out!), to mention entropy (whose existence and definition should *follow from* the second law), or to comment that some carefully described, rather esoteric process is impossible. He has settled, therefore, on the statement that *the macroscopic properties of any isolated system eventually assume constant values*. Isolated systems evolve until they reach equilibrium. Perpetual motion in an isolated system is impossible. Certainly this statement is accessible to our day-to-day experience. We sense it every time we get hungry, and when we look in the mirror each morning the irreversibility of nature is forcibly brought home to us. Furthermore, one can quickly derive the conventional statements of the zeroth and second laws by invoking the empirically observed properties of heat engines.* An interesting question, though, is to what extent this statement of the second law is actually true.

Let us raise the issue by means of an example. Suppose two bottles of equal size are connected by a stopcock. Initially, the stopcock is closed, one bottle contains a dilute gas with N molecules, and the other bottle is empty. At time zero we remove the constraint by opening the stopcock. The entropy, viewed as a function of N, E, and constraints (V), jumps discontinuously to its new value with $\Delta S = Nk\ln 2$, as given by thermodynamics. Or, if one views entropy as a measure of information, opening the stopcock destroys one bit of information about each particle (in which of the two bottles it was located), which means that $\Delta S = Nk\ln 2$, by Eq. 7-27. Of course local properties such as pressure and density, which one would calculate using the entropy $S = S(E, V, N)$ as a thermodynamic potential,† do not jump discontinuously to their new equilibrium values. Instead, they approach equilibrium at a rate that cannot be obtained from equilibrium theory. *Nonequilibrium* statistical mechanics or thermodynamics must be used to predict how fast the new equilibrium will be reached. Only after the new equilibrium is established, can the new value of entropy be used to predict the result of any experiment. After opening the stopcock, it takes a while for the initial information to become worthless in making predictions—the system has inertia; it remembers its initial state over the period of its relaxation time.

In 1890 Henri Poincaré threw a wrench into the orderly development of the molecular theory of the approach to equilibrium by proving that no matter how wild and unlikely a departure from the equilibrium state one imagines, such a fluctuation *will occur* if only one waits long enough. Thus our statement of the second law is false, and given long enough, all N molecules will find themselves back in the first bottle! Thus we must consider how false our statement of the second law really is.

We are not talking here about uncertainty in measuring; we suppose that measurements on the system give the exact value of the property being measured. The topic of experimental error is a branch of applied probability theory different from statistical mechanics. Ensemble averages are the values predicted by statistical mechanics for the various properties. However, most properties need not have the ensemble average value; when some property deviates from its mean value (deviation from the mean defined by Eq. 2-30), this property is said to be in a **fluctuation**. It is

*Andrews, Chaps. 8–11.

†Andrews, Chap. 13 and Appendix B.

impossible to predict *when* an equilibrium system will undergo a fluctuation because the necessary information (about the positions and velocities, say, of a group of particles) is not available. However, one *can* predict that the probabilities of various fluctuations equal the fraction of members of the ensemble exhibiting such fluctuations. The likelihood of significant fluctuations is a measure of the usefulness of the predictions of statistical mechanics. It is far different to predict the value 2.00 for a result when the ensemble members on the one hand have the values 1.98, 2.00, 2.00, 2.01 2.01, and, on the other hand, have the values 0.10, 0.50, 1.50, 3.50, 4.40.

What about Poincaré's proof that all the molecules will come back into the first bottle? If the probability that any one gas molecule be in the first bottle is $\frac{1}{2}$, the probability that all N will be there is $(\frac{1}{2})^N$. If N is 10, this probability is about 10^{-3}, and the fluctuation would probably be seen about every thousandth time one looked at the system. However, if N is 100, the probability that all were in the first bottle is about 10^{-30}; thus if one looked at the bottles once a second for 10^{12} times as long as the estimated age of the universe, he might expeect to see the fluctuation! Therefore, the second law as we have stated it is indeed false, even on the scale of macroscopic objects. However, on the time scale of our universe, which is what we are stuck with, it is true.

If macroscopic fluctuations do not occur, what kind do? We have already seen in the discussion of Eqs. 5-18 to 5-24 and Fig. 5-3 that significant fluctuations in the energy of a system in equilibrium with a heat bath have utterly negligible likelihood. The root mean square deviation from the mean for the system energy was calculated to be approximately $N^{-1/2}$ times the average energy itself. True, the system *could* have any energy. But the density of states (which rapidly increases with energy) combines with the probability of a particular state (which goes down as $e^{-\beta E_{\text{state}}}$) to make one particular energy overwhelmingly likely.

THE GRAND CANONICAL ENSEMBLE

Let us study fluctuations in number of molecules present in a region at equilibrium much as we treated energies in Chap. 5. To do this, an ensemble must be constructed in which N is a variable. This is what J. Willard Gibbs called a **grand ensemble**, in contrast to a **petit ensemble**, in which each member has the same number of particles. In a grand ensemble not only do members exist in each allowed quantum state for a fixed N, but there are members in each allowed state for each possible value of N. The probability P_{Nj} that the system will be found containing N particles in the jth N-particle quantum state is the fraction of members of the ensemble containing just N particles in the quantum state j.

How shall we construct the N-dependence of the distribution function? Let the system be a well-defined volume V which is open to the exchange of particles with its surroundings. We view it as being *in equilibrium with a particle bath*, much as the system in a canonical ensemble is in equilibrium with a heat bath. In fact, Fig. 5-1 can be modified as shown in Fig. 16-1 to represent the system in material equilibrium with a particle bath, and both system and particle bath in thermal equilibrium with a heat bath. We view the system plus particle bath as a *composite* containing $\mathcal{N}$ particles in a volume $\mathcal{V}$ and being in equilibrium with a heat bath. Thus the composite can be represented by a canonical distribution:

$$P_{\mathcal{N}}(i) = \frac{e^{-\beta \mathcal{E}_i}}{Q_{\mathcal{N}}(\mathcal{V})}. \qquad \text{(system plus particle bath)} \qquad (16\text{-}1)$$

In Eq. 16-1, $P_{\mathcal{N}}(i)$ is the probability that the $\mathcal{N}$-particle composite is in its ith quantum state, which has energy $\mathcal{E}_i$. The distinction between system and particle bath must be clear enough that the following distinctions are meaningful: there are N particles in the system and $\mathcal{N} - N$ in the bath; the N particles of the system are in quantum state j, which has the system energy E_j; the $\mathcal{N} - N$ particles of the bath are in quantum state k of energy $\mathcal{E}_k$, where $\mathcal{E}_i = E_j + \mathcal{E}_k$. The subscript on Q refers to the number of particles in the composite system plus particle bath.

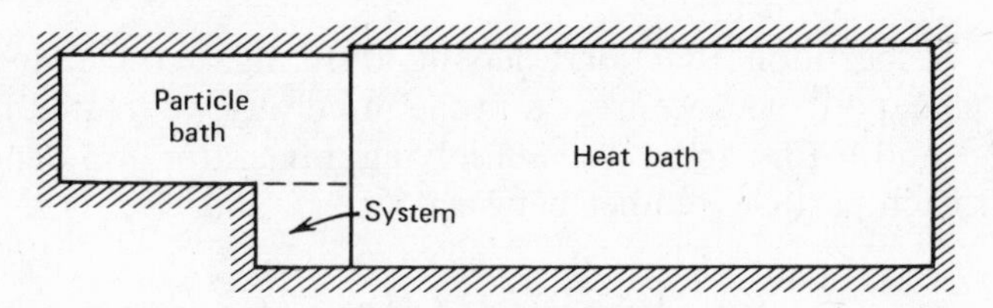

Fig. 16-1. System and particle bath in equilibrium through a wall open to passage of material; both areas in equilibrium with heat bath through a diathermal wall.

The probability $P_{N,j}$ that the system contains N particles in state j is the sum of terms of the form of Eq. 16-1 over all states k of the composite showing N particles to be in the system in state j. It makes no difference what state the $\mathcal{N} - N$ particles of the bath are in; they all contribute to $P_{N,j}$; and because of the heat bath, all values of $\mathcal{E}_k$ are possible. Thus we have

$$P_{N,j} = \sum_k \frac{e^{-\beta \mathcal{E}_i}}{Q_{\mathcal{N}}(\mathcal{V})} = e^{-\beta E_j} \frac{\Sigma_k e^{-\beta \mathcal{E}_k}}{Q_{\mathcal{N}}(\mathcal{V})} = e^{-\beta E_j} \frac{Q_{\mathcal{N}-N}(\mathcal{V} - V)}{Q_{\mathcal{N}}(\mathcal{V})}. \qquad (16\text{-}2)$$

The partition functions in the ratio of Eq. 16-2 have different particle numbers and different volumes, although $N \ll \mathcal{N}$ and $V \ll \mathcal{V}$. We use the V-dependence of Q, given in Eq. 8-24, to change $Q(\mathcal{V}-V)$ in Eq. 16-2 to $Q(\mathcal{V})$:

$$\left(\frac{\partial \ln Q}{\partial V}\right)_T = \beta P. \tag{16-3}$$

Thus we can put

$$\ln Q(\mathcal{V}-V) = \ln Q(\mathcal{V}) + \int_{\mathcal{V}}^{\mathcal{V}-V} \left(\frac{\partial \ln Q}{\partial V'}\right)_T dV' \tag{16-4}$$

$$\ln Q(\mathcal{V}-V) = \ln Q(\mathcal{V}) + \int_{\mathcal{V}}^{\mathcal{V}-V} \beta P \, dV' = \ln Q(\mathcal{V}) + \beta PV, \tag{16-5}$$

since V is very tiny compared with $\mathcal{V}$, and P therefore does not change appreciably during the integration. Thus we have proved

$$Q(\mathcal{V}-V) = Q(\mathcal{V}) e^{-\beta PV}, \tag{16-6}$$

which can be used in Eq. 16-2 to yield

$$P_{N,j} = e^{-\beta PV - \beta E_j} \frac{Q_{\mathcal{N}-N}(\mathcal{V})}{Q_{\mathcal{N}}(\mathcal{V})}. \tag{16-7}$$

This ratio of partition functions, having differing particle numbers, is familar from Eq. 9-8, and we treat it in the same way we treated Eq. 9-8 to get Eqs. 9-9 and 9-10, namely by multiplying numerator and denominator by a Q for each particle number between $\mathcal{N}-N$ and $\mathcal{N}$:

$$P_{N,j} = e^{-\beta PV - \beta E_j} \frac{Q_{\mathcal{N}-1}}{Q_{\mathcal{N}}} \frac{Q_{\mathcal{N}-2}}{Q_{\mathcal{N}-1}} \frac{Q_{\mathcal{N}-3}}{Q_{\mathcal{N}-2}} \cdots \frac{Q_{\mathcal{N}-N}}{Q_{\mathcal{N}-V+1}}. \tag{16-8}$$

Since $N \ll \mathcal{N}$, the N ratios of partition functions in Eq. 16-8 all have approximately the same value, which we know from Eq. 8-20 or Eq. 9-5 to be $e^{\beta\mu}$. Our final result is

$$\boxed{P_{N,j} = e^{-\beta PV + \beta N\mu - \beta E_j} \equiv \frac{1}{\Xi} e^{\beta N\mu - \beta E_j},} \tag{16-9}$$

where we have given the reciprocal of the normalization constant $e^{-\beta PV}$ the symbol Ξ. The value of Ξ is found by normalization in the same manner used to find Q:

$$\Xi = \sum_N \sum_j e^{\beta N\mu - \beta E_j} = e^{\beta PV}. \tag{16-10}$$

The quantity Ξ is called the **grand partition function**, and an ensemble constructed according to Eq. 16-9 is said to be a **grand canonical ensemble**. The "grand" means that N is explicitly allowed to vary; "canonical" indicates that the members of the ensemble for any fixed value of N form a canonical ensemble among themselves. The grand canonical ensemble can be viewed as a collection of canonical ensembles, one for each number of particles, weighted according to Eq. 16-9. This view recalls that in Chap. 5 we presented a canonical ensemble as a collection of microcanonical ensembles of differing energies. If there are several different species of molecule in the system, the product $N\mu$ in Eqs. 16-9 and 16-10 will be replaced by a sum over the $N_i\mu_i$ products for each species.

FLUCTUATIONS IN DENSITY

Density fluctuations can be studied by using the grand canonical ensemble to determine the mean square deviation of the number of particles in some chosen volume V. This procedure is completely analogous to the approach used in relation to energy in Eqs. 5-18 to 5-23. The average number of particles $\overline{N}$ in the volume V is

$$\overline{N} = \frac{\sum_N \sum_j e^{\beta N\mu - \beta E_j} N}{\sum_N \sum_j e^{\beta N\mu - \beta E_j}}, \tag{16-11}$$

from Eq. 16-9. The mean square deviation in N is given by Eq. 2-40:

$$\overline{\delta N^2} = \overline{N^2} - \overline{N}^2. \tag{16-12}$$

We can express the right-hand side of Eq. 16-12 for our system by a trick: differentiating Eq. 16-11 with respect to μ at constant T and V yields

$$\frac{\partial \overline{N}}{\partial \mu} = \frac{\beta \sum_N \sum_j e^{\beta N\mu - \beta E_j} N^2}{\sum_N \sum_j e^{\beta N\mu - \beta E_j}} - \frac{\left(\sum_N \sum_j e^{\beta N\mu - \beta E_j} N\right)^2}{\left(\sum_N \sum_j e^{\beta N\mu - \beta E_j}\right)^2}. \tag{16-13}$$

These two terms can be identified immediately:

$$\frac{\partial \overline{N}}{\partial \mu} = \beta(\overline{N^2} - \overline{N}^2) = \beta\,\overline{\delta N^2}\,, \tag{16-14}$$

$$\boxed{\overline{\delta N^2} = kT\left(\frac{\partial N}{\partial \mu}\right)_{T,V} = \frac{kT}{(\partial \mu/\partial N)_{T,V}}.} \tag{16-15}$$

This result is analogous to Eqs. 5-21 or 5-23 for energy fluctuations. The relative fluctuation in N is easily found for an ideal gas by using Eq. 9-30:

$$\mu = kT\ln\frac{N}{q}\,; \qquad \left(\frac{\partial \mu}{\partial N}\right)_{T,V} = \frac{kT}{N}\,; \qquad \overline{\delta N^2} = N. \tag{16-16}$$

Therefore, exactly as with energies, the ratio of $(\overline{\delta N^2})^{1/2}$ to $\overline{N}$ is $1/\sqrt{N}$.

Since particle density ρ is just N/V, the same conclusion also holds for density fluctuations: when density is measured by looking in a volume containing an average of N particles, the ratio of $(\overline{\delta\rho^2})^{1/2}$ to $\overline{\rho}$ is $N^{-1/2}$. Thus relative fluctuations in density can be made as large as desired by looking in smaller and smaller volumes. As expected, therefore, the uniformity of density observed in a system depends on the fineness of the scale employed. That density fluctuations indeed exist in the air is testified to by the blue sky. If there were no fluctuations in density, sunlight would not be appreciably scattered by the atmosphere; it would go in straight lines—as it does, for example, in the vacuum over the surface of the moon—and the sky would be nearly black. Small adjacent regions of greatly different densities act as scattering centers for light, however, deflecting it in all directions. A significant density fluctuation must occur over a region whose dimensions are of the order of the wavelength of the light before there is appreciable scattering. Since blue light has shorter wavelength than red, and since significant fluctuations are more likely in small regions than large, blue light is preferentially scattered over red. Thus the light that goes on straight is reddened as the blue is scattered out of it. This is a partial explanation of the red color of sunsets and sunrises, where one is looking into the sun along a path that has gone through the upper atmosphere for a long distance.

The conclusion we reach here is the same one drawn following Eq. 5-24. A particle bath is a source or sink for particles characterized by a constant value of the chemical potential. One may, if he wishes, use a grand canonical ensemble to represent a macroscopic system, regardless of whether the system is in equilibrium with a particle bath. This is because significant fluctuations in N are negligible. This procedure is often followed in statistical mechanics, to simplify various procedures.

It is interesting to note what happens to Eq. 16-15 in the event of a first-order phase transition, such as from liquid to vapor. When both phases coexist, the number of particles can be changed with no change in μ as one simply increases the relative amount of one phase and decreases the other. Thus $(\partial N/\partial\mu)$ is infinite, as is $\overline{\delta N^2}$ by Eq. 16-15. Thus we would expect $\overline{N}$ to be a poor prediction of the number of particles in V in this case, and indeed it is. The average density could lie anywhere between the densities of the two pure phases.

The existence of fluctuations and their agreement with theory is confirmed by studying the Brownian motion or its equivalent, present in all systems at temperatures above 0 K. Random thermal noise in transmission lines, for example, limits the signal which such lines can carry. In a quiet enough place, one can even hear the Brownian motion of his own eardrums, caused by their irregular bombardment by air molecules. Fluctuations are transitory, unforecastable though statistically predictable violations of any statement of the second law that refers to observable phenomena (e.g., our own regarding the impossibility of perpetual motion, Clausius', Kelvin's, Caratheodory's). However, they do not represent violations of the second law if that law merely talks about the entropy as a mathematical function. In that event, fluctuations simply destroy one's ability to calculate observable quantities from the entropy.*

TEMPERATURE: POSITIVE AND NEGATIVE

We have not mentioned fluctuations in either temperature or entropy because these quantities are statistical, not mechanical. Entropy is a measure of the information one has about which quantum state an equilibrium system is in. Temperature measures the steepness of the exponentially damped probability $e^{(A-E_i)/kT}$ that the system will be in quantum state i of energy E_i. As such it is a property of the heat bath in equilibrium with the system and does not fluctuate.

However, one usually does not measure energy but instead measures temperature, using as a thermometer an instrument that interacts not with the whole system but with only a part of it. Thus it might be useful to change the definition of T, to give it mechanical meaning. Since the energy

*An interesting discussion of fluctuations is given in R. P. Feynman, R. B. Leighton, and M. Sands, *The Feynman Lectures on Physics*, Vol. 1 (Addison-Wesley Publishing Co., Reading, Mass), Sec. 46, See also F. C. Andrews, *Proc. Nat. Acad. Sci.* (U.S.), **54**, 13 (1965).

of a macroscopic system fluctuates so little, one might use statistical mechanics to calculate the average energy as a function of temperature, $E = E(T, V, N)$. Then this equation could be inverted to yield $T = T(E, V, N)$. Then, as E fluctuated, T would fluctuate also. In this case, if N is made smaller, the fluctuations in T become much more pronounced. Thus since the average kinetic energy of a single particle is $\frac{3}{2}kT$, one could define temperature as $2/3k$ times the kinetic energy of some particle under consideration. Temperature so defined would fluctuate wildly, but for certain purposes the definition might prove to be useful. We shall consider the behavior of individual particles at greater length in the last part of the book.

One interesting curiosity arising from the statistical definition of temperature is the concept of negative temperatures. Consider a solid whose particles possess permanent magnetic dipoles of strength $\boldsymbol{\mu}$. In a magnetic field $\mathcal{H}$ the energy of each dipole relative to the field is $\boldsymbol{\mu} \cdot \mathcal{H}$, and the probability that a single particle has the angle between $\boldsymbol{\mu}$ and $\mathcal{H}$ given by θ is proportional to $e^{-\beta \boldsymbol{\mu} \cdot \mathcal{H}} = e^{-\beta \mu \mathcal{H} \cos\theta}$. At very low temperatures, approaching 0 K, this exponential damping assures that the dipoles are almost perfectly aligned in opposition to the field, making θ very nearly 180° for all particles, which in turn makes $\cos\theta$ nearly -1, the smallest value it can possibly have. This is shown in Fig. 16-2a. All the particles do not have $\theta = 180°$ only because of the thermal energy that keeps jostling them out of their preferred alignment. It is possible to flip the direction of the applied field, to change it very quickly by 180°. The particles are now characterized by a distribution proportional to $e^{\beta \boldsymbol{\mu} \cdot \mathcal{H}}$ (Fig. 16-2*b*). They have more energy than they would have at equilibrium, but the only way they can eliminate the excess is to give it up to vibrational modes in the crystal. It may take many seconds for this equilibration or relaxation (Figs. 16-2*c* and 16-2*d*) to occur, since the process is inefficient. If one examined just the magnetic part of the system on a time scale that was short compared with the relaxation time, one could interpret the distribution of dipoles (Fig. 16-2*b*) as equilibrium, but with β negative. The temperature would be negative. Of course, this is really a nonequilibrium condition, but the coupling of the magnetic to the vibrational degrees of freedom is often so weak that negative temperatures exist for some time. As the excess magnetic energy is given up to the lattice, the magnetic dipoles "cool down," and β for the magnetic system is observed to increase from its initial negative value, pass through zero (Fig. 16-2*c*), and continue growing until it reaches the β of the lattice (Fig. 16-2*d*). Thus as the magnetic dipoles lose energy and cool off, the magnetic temperature changes from its initial negative value (Fig. 16-2*b*) to $-\infty$ (Fig. 16-2*c*), which is the same as $+\infty$, then decreases further until it reaches the temperature of the lattice (Fig. 16-2*d*). Thus the highest possible temperature is -0 K (really the limit as $\Theta \rightarrow 0$ of $-\Theta$ K); the lowest temperature is of course $+0$ K (or the limit as $\Theta \rightarrow 0$ of Θ K).

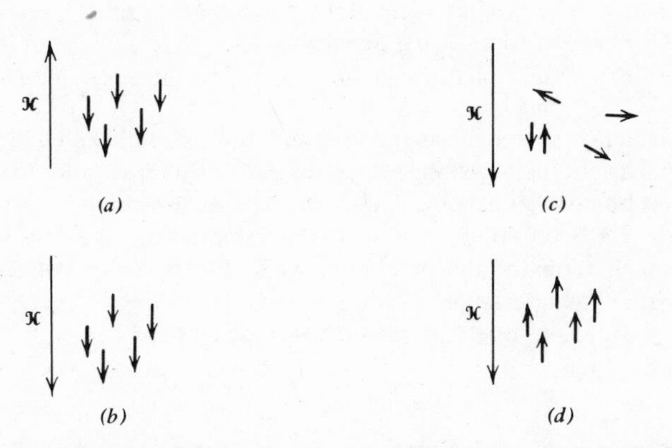

Fig. Fig. 16-2. (*a*) Magnetic dipoles aligned against field at T near $+0$. (*b*) Field direction suddenly reversed; dipoles have T near -0. (*c*) Dipoles have lost about half their excess energy to lattice, have T near $-\infty = +\infty$. (*d*) Dipoles finally regain thermal equilibrium with lattice at T near $+0$.

The thermodynamic formula

$$k\beta = \left(\frac{\partial S}{\partial E} \right)_V \tag{16-17}$$

sheds light on the meaning of a negative value of β. The value of S is minimized whenever the dipoles are all *aligned*, regardless of whether they are aligned with or against the field. In the negative temperature region, an increase in magnetic energy tends to align the dipoles against the field, thus ordering the system and decreasing the entropy. Heat always flows from regions of low β to regions of high β. Of course negative β can exist only for such features of a nonequilibrium system as magnetic dipoles, whose entropy varies with energy such that Eq. 16-17 can assume negative values.

PROBLEMS

16-1. In what volume of an ideal gas at 1 atm and 300 K is $\overline{N} = 1.00$? For so small a volume, how large is $\overline{(\delta N^2)}^{1/2}$?

16-2. Contrast $\overline{(\delta F^2)}^{1/2}$ with $\left[\overline{(\delta F)}^2\right]^{1/2}$, where F is an arbitrary function, stating the meaning of each in words. Which is larger? Does the latter quantity differ from $\overline{\delta F}$?

16-3. At temperatures very small compared with Θ_D, Eq. 13-34 for the energy of a Debye crystal becomes $E = 3\pi^4 NkT^4/5\Theta_D{}^3$. Express $(\overline{\delta E^2}/\overline{E}^2)$ as a function of N, T, and Θ_D. In thermodynamics it is customary to consider energy and temperature to be simultaneously determinable quantities, the one determining the other theoretically. Below what temperature does $(\overline{\delta E^2}/\overline{E}^2)^{1/2}$ exceed unity for one milligram of carbon (diamond)? See Table 13-1.

16-4. Another measure of the average of $E - \overline{E}$ in the ensemble is $\overline{\delta E^3}$. Prove that $\overline{\delta E^3} = -\partial^3 \ln Q/\partial\beta^3$.

16-5. Consider a system consisting of liquid and its equilibrium vapor sealed in a cylinder by a frictionless piston, whose weight in the gravitational field exactly balances the vapor pressure. Above the piston is an evacuated volume, and the entire cylinder is a fixed volume. What is the value of $\overline{\delta E^2}/\overline{E}^2$ and why? What does this mean in terms of the prediction we would make of the energy of this system, knowing its temperature?

16-6. Discuss density fluctuations in two-phase systems.

PART 4

SEMICLASSICAL STATISTICAL MECHANICS

17 DISTRIBUTION FUNCTIONS

THE COMPLETE DISTRIBUTION FUNCTION

One of the great advantages of the atomic-molecular model of physical reality is the useful intuitive picture it affords of what is going on at the submicroscopic level to "cause" macroscopic phenomena. Unfortunately, the fully quantum mechanical treatment of the center of mass motion as given in Chap. 10 is most unintuitive. A classical mechanical treatment that preserved such concepts as molecular positions, velocities, trajectories, and collisions would be much more helpful in this respect. Furthermore, the quantum treatment is almost useless whenever the density of the fluid being studied is so large that the ideal gas approximation breaks down. In this case, too, a classical treatment of the center of mass motion is needed. In addition, when one considers non-equilibrium statistical mechanics, the classical picture is much more useful than the purely quantum mechanical.

In the semiclassical treatment of statistical mechanics, only the centers of mass of the molecules are handled classically. Their internal behavior is treated by the quantum methods of Chap. 11, and we shall ignore the internal problem in this part of the book, focusing only on the centers of mass, each of which is characterized by a position **r** and a momentum **p** (or velocity $\mathbf{v}=\mathbf{p}/m$). The reader is referred to the classical mechanics section of Chap. 3 for a review of the notation used.

For the reasons discussed in Chap. 5, if the system is in equilibrium with a heat bath, the result is a canonical ensemble. The complete N-particle distribution function takes the form of an exponential in the total energy associated with the centers of mass:

$$f_N(\mathbf{r}^N,\mathbf{p}^N)=ae^{-\beta E(\mathbf{r}^N,\mathbf{p}^N)}, \qquad (17\text{-}1)$$

where a is the normalization constant. If we define f_N as in Chap. 3, the N particles are considered to be truly classical mechanical (i.e., distinguishable); each one effectively has a number painted on its back. Then the definition of Chap. 3 is meaningful: $f_N\, d\mathbf{r}^N\, d\mathbf{p}^N$ is the joint probability that particle 1 is in $d\mathbf{r}_1\, d\mathbf{p}_1$ and particle 2 in $d\mathbf{r}_2\, d\mathbf{p}_2, \ldots$. Of course a proper semiclassical treatment of a fluid must include the quantum mechanical result of Chap. 9 for indistinguishable particles: particles cannot have labels; interchange of two particles does not lead to a new state. We shall keep that in mind and use the following **definition of** f_N, which is correct for a semiclassical treatment: $f_N(\mathbf{r}^N, \mathbf{p}^N)\, d\mathbf{r}^N\, d\mathbf{p}^N$ is the joint probability that some particle has its x-coordinate between x_1 and $x_1 + dx_1$, its y-coordinate between y_1 and $y_1 + dy_1$, its z-coordinate between z_1 and $z_1 + dz_1$ (in other words, its position lies within the volume element $d\mathbf{r}_1$ located at the point $\mathbf{r}_1$), *and* its momentum lies within $d\mathbf{p}_1$ about the point $\mathbf{p}_1$; *and* that another particle lies in $d\mathbf{r}_2$ at $\mathbf{r}_2$ and has its momentum in $d\mathbf{p}_2$ at $\mathbf{p}_2$; *and* that another particle lies in $d\mathbf{r}_3$ at $\mathbf{r}_3$ and has its momentum in $d\mathbf{p}_3$ at $\mathbf{p}_3; \ldots$, etc., for N different combinations of volume and momentum elements. Since the likelihood of multiple occupancy is negligible (Boltzmann statistics are valid), the semiclassical definition of f_N is just $N!$ times the classical definition. Since the **r**'s and **p**'s used in Eq. 17-1 refer to positions and momenta of numbered particles, f_N in the classical definition is normalized to unity, each of the N particles requiring *some* position and momentum. Thus f_N in the semiclassical definition is normalized to $N!$ Readers must always ascertain how a given author defines f_N; if the author does not specify, his normalization will give it away. We shall use the semiclassical definition; therefore the particles are truly indistinguishable and f_N is normalized to $N!$:

$$a^{-1} = \frac{1}{N!} \int d\mathbf{r}^N\, d\mathbf{p}^N e^{-\beta E(\mathbf{r}^N, \mathbf{p}^N)}. \tag{17-2}$$

The integrals over each position **r** range over the volume V; the integrals over each component of momentum range from $-\infty$ to $+\infty$. The integral in Eq. 17-2 is sometimes called the **phase integral** because it is over the classical phase space of the N particles.

The energy in Eqs. 17-1 and 17-2 is the sum of the kinetic energy of each particle plus the intermolecular potential energy:

$$E = \sum_{i=1}^{N} \frac{p_i^2}{2m} + U(\mathbf{r}^N). \tag{17-3}$$

We have taken the intermolecular potential energy U to depend on only the positions of the N particles. This is of course an approximation, and a poor one for nonspherical molecules. For them the potential energy depends not only on center of mass positions but also on the angles of alignment of the molecules with respect to one another. Whenever we use

Eq. 17-3, therefore, we are making a spherical approximation to the true shapes of the molecules.

Use of Eq. 17-3 in Eqs. 17-1 and 17-2 gives the complete N-particle distribution function for the centers of mass of the molecules in a fluid at equilibrium:

$$f_N(\mathbf{r}^N,\mathbf{p}^N)=\frac{N!\exp\left[-\beta\sum_{i=1}^{N}(p_i^2/2m)-\beta U(\mathbf{r}^N)\right]}{\int d\mathbf{p}^N\exp\left(-\beta\sum_{i=1}^{N}(p_i^2/2m)\right)\int d\mathbf{r}^N e^{-\beta U(\mathbf{r}^N)}}. \tag{17-4}$$

We need carry f_N no further here; its use is especially important in the treatment of dense gases and liquids and in nonequilibrium statistical mechanics.

REDUCED DISTRIBUTION FUNCTIONS

A variety of reduced distribution functions are useful in working with fluids. We define, calculate, and discuss a few of them here. The first is the **one particle or singlet distribution function** f_1, defined as follows:

$$f_1(x,y,z,p_x,p_y,p_z)dx\,dy\,dz\,dp_x\,dp_y\,dp_z\equiv f_1(\mathbf{r},\mathbf{p})d\mathbf{r}\,d\mathbf{p} \tag{17-5}$$

is the probability that a particle (not a *particular* particle, since particular particles cannot be distinguished, but *any* particle in the system) has its x-coordinate between x and $x+dx$, its y-coordinate between y and $y+dy$, its z-coordinate between z and $z+dz$, its x-component of momentum between p_x and p_x+dp_x, its y-component of momentum between p_y and p_y+dp_y, and its z-component of momentum between p_z and p_z+dp_z. Or, more briefly, Eq. 17-5 represents the probability that there be a particle in the elements $d\mathbf{r}$ at $\mathbf{r}$ and $d\mathbf{p}$ at $\mathbf{p}$.

The meaning of these abstract definitions is perhaps clarified by Fig. 17-1, illustrating the configuration and momentum spaces for a particle. Together these spaces form the complete phase space of the particle. Configuration space is limited to the interior of a box of volume V. Momentum space, however, is unlimited. Vectors $\mathbf{r}$ and $\mathbf{p}$ have been drawn arbitrarily, and at their ends, volume elements $d\mathbf{r}$ and $d\mathbf{p}$ have been indicated. In each member of the ensemble all N particles have definite positions and momenta. For a given choice of $\mathbf{r},\mathbf{p},d\mathbf{r}$, and $d\mathbf{p}$, one can simply ascertain what fraction of the members has a particle inside the volume element $d\mathbf{r}$ whose momentum vector ends inside the element $d\mathbf{p}$.

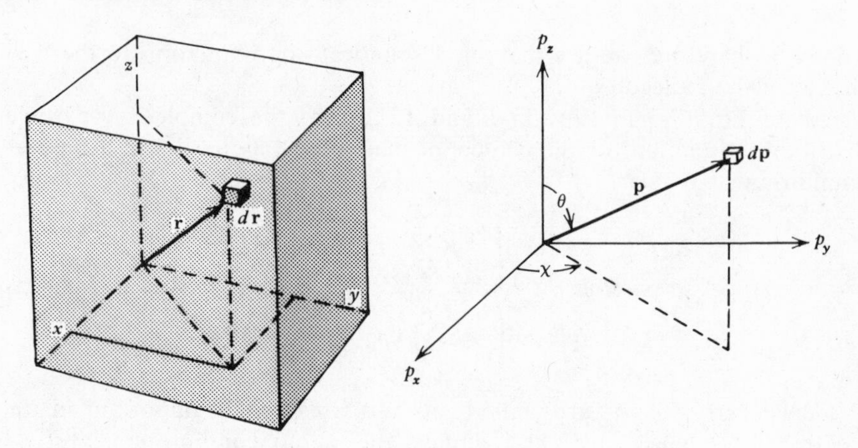

Fig. 17-1. (*a*) Configuration space and (*b*) momentum space of a particle.

This fraction is the probability, Eq. 17-5, for that choice of $\mathbf{r}, \mathbf{p}, d\mathbf{r}$, and $d\mathbf{p}$. The ensemble therefore defines the one particle distribution function $f_1(\mathbf{r}, \mathbf{p})$.

The singlet distribution function is obtained by reducing f_N, that is, by integrating over the $N-1$ positions and momenta that are not of interest. Then we divide the result by $(N-1)!$, since f_1 must be normalized to N, which is the total number of particles that will be found somewhere in one-particle phase space. All the dependence on $\mathbf{r}_2, \mathbf{r}_3, \ldots, \mathbf{r}_N, \mathbf{p}_2, \mathbf{p}_3, \ldots, \mathbf{p}_N$ (i.e., on $\mathbf{r}^{N-1}$ and $\mathbf{p}^{N-1}$) will thus be integrated out of f_N in the process of obtaining f_1. The result will be a function proportional to the exponential of the total energy of a single particle:

$$f_1(\mathbf{r}_1, \mathbf{p}_1) \equiv \frac{1}{(N-1)!} \int d\mathbf{r}^{N-1}\, d\mathbf{p}^{N-1} f_N(\mathbf{r}^N, \mathbf{p}^N), \qquad \text{(normalized to } N) \qquad (17\text{-}6)$$

$$f_1(\mathbf{r}, \mathbf{p}) = A e^{-\beta\epsilon(\mathbf{r}, \mathbf{p})}. \qquad (17\text{-}7)$$

This is analogous to the Boltzmann factor of the quantum treatment, Eq. 9-14. The energy $\epsilon(\mathbf{r}, \mathbf{p})$ is the average energy of a single particle in the fluid at the point $\mathbf{r}$, having the momentum $\mathbf{p}$. If there is no external field such as gravity (treated in Chap. 19), acting on the particles, the position of the particle in the vessel does not affect its average energy. Since its average intermolecular potential energy is also constant, independent of its position, we can write the result

$$f_1(\mathbf{r}, \mathbf{p}) = \alpha e^{-\beta p^2/2m}. \qquad \text{(no external fields)} \qquad (17\text{-}8)$$

The value of α may be found by normalization:

$$\int d\mathbf{r} \int d\mathbf{p} f_1(\mathbf{r},\mathbf{p}) = N = \alpha \int d\mathbf{r} \int d\mathbf{p}\, e^{-\beta p^2/2m}, \tag{17-9}$$

$$\alpha^{-1} = \frac{V}{N} \int dp_x\, dp_y\, dp_z\, e^{-\beta(p_x^2+p_y^2+p_z^2)/2m}, \tag{17-10}$$

$$\alpha^{-1} = \frac{V}{N}\left[\int_{-\infty}^{\infty} dz\, e^{-\beta z^2/2m}\right]^3 = \frac{V}{N}\left(\frac{2\pi m}{\beta}\right)^{3/2}. \tag{17-11}$$

$$\boxed{f_1(\mathbf{r},\mathbf{p}) = \frac{N}{V}\left(\frac{\beta}{2\pi m}\right)^{3/2} e^{-\beta p^2/2m} = \rho\left(\frac{\beta}{2\pi m}\right)^{3/2} e^{-\beta p^2/2m}.} \tag{17-12}$$

The value of the integral was obtained from Eq. B-1 in Appendix B. Note the factorization of f_1: the total number of particles one expects to find in $d\mathbf{r}$ is $\rho\, d\mathbf{r}$, and the fraction of particles with momentum in $d\mathbf{p}$ is $(\beta/2\pi m)^{3/2} e^{-\beta p^2/2m}$. There is no correlation between position and momentum in this joint probability amplitude.

The second reduced distribution function we mention is the rather trivial example of the **one-particle or singlet-density** $\rho(\mathbf{r})$, which is defined as follows: $\rho(\mathbf{r})\, d\mathbf{r}$ is the probability of finding a particle in the volume element $d\mathbf{r}$ about the point $\mathbf{r}$. Thus ρ is normalized to N, the total number of particles that must be found somewhere. One obtains $\rho(\mathbf{r})$ from $f_1(\mathbf{r},\mathbf{p})$ by integrating f_1 over all $\mathbf{p}$, since we do not care what momentum the particle has:

$$\rho(\mathbf{r}) \equiv \int d\mathbf{p} f_1(\mathbf{r},\mathbf{p}). \qquad \text{(normalized to } N\text{)} \tag{17-13}$$

At equilibrium in the absence of external fields, ρ is independent of $\mathbf{r}$ and must be

$$\boxed{\rho = \frac{N}{V}.} \qquad \text{(no external fields)} \tag{17-14}$$

The third reduced distribution function we mention is the **one-particle momentum distribution function** $\varphi_1(\mathbf{p})$, which is defined as follows: $\varphi_1(\mathbf{p})\, d\mathbf{p}$ is the probability that a particle in the system has its momentum within $d\mathbf{p}$ about $\mathbf{p}$. Therefore, φ_1 is normalized to unity, since each particle has *some* value for its momentum. The integration of f_1 over all $\mathbf{r}$ gives N times φ_1:

$$\varphi_1(\mathbf{p}) = \frac{1}{N} \int d\mathbf{r} f_1(\mathbf{r},\mathbf{p}). \qquad \text{(normalized to 1)} \tag{17-15}$$

Clearly, at equilibrium we have

$$f_1(\mathbf{r},\mathbf{p})=\rho(\mathbf{r})\varphi_1(\mathbf{p}), \tag{17-16}$$

and therefore

$$\boxed{\varphi_1(\mathbf{p})=(2\pi mkT)^{-3/2}e^{-p^2/2mkT}.} \tag{17-17}$$

Since momenta are less easily visualized than velocities, we introduce a fourth reduced distribution function, the **one-particle velocity distribution function** $\Phi_1(\mathbf{v})$, defined as follows: $\Phi_1(\mathbf{v})\,d\mathbf{v}$ is the probability that a particle in the system has its velocity in the range $d\mathbf{v}$ about $\mathbf{v}$. For nonrelativistic speeds, $\mathbf{p}=m\mathbf{v}$; thus by equating equivalent probabilities, we can find Φ_1:

$$\Phi_1(\mathbf{v})\,d\mathbf{v}=\varphi_1(\mathbf{p})\,d\mathbf{p}, \tag{17-18}$$

$$\Phi_1(\mathbf{v})\,d\mathbf{v}=\varphi_1(\mathbf{p})m^3\,d\mathbf{v}, \tag{17-19}$$

$$\boxed{\Phi_1(\mathbf{v})=\left(\frac{m}{2\pi kT}\right)^{3/2}e^{-mv^2/2kT}.} \tag{17-20}$$

The m^3 appears because $dp_x=m\,dv_x$, and $d\mathbf{p}=dp_x\,dp_y\,dp_z$. The reader may quickly verify the normalization of Eq. 17-20. The result, Eq. 17-20, is called the **Maxwell velocity distribution.**

The Maxwell distribution gives the probabilities of different velocities of molecules in the fluid. Since N is so large, the Maxwell distribution is an extraordinarily accurate prediction of the way the actual velocities of the N molecules in the system are distributed. Of course, we do not *know* what the distribution of velocities in the system is; Eq. 17-20 reflects the ensemble and is only a prediction or expectation about the system. However, significant departures from Eq. 17-20 for the distribution of particle velocities in a macroscopic equilibrium system are highly unlikely. The Maxwell distribution has been verified to the limit of experimental accuracy in an enormous number of experiments.

It is interesting to view the Maxwell distribution graphically. We look at $\Phi(v_x)$, the **distribution function for the** x**-component of velocity**, defined thus: $\Phi(v_x)\,dv_x$ is the probability that the x-component of velocity of a particle in the system lies between v_x and v_x+dv_x. It is clear from examining Eq. 17-20 that $\Phi_1(\mathbf{v})$ factors into $\Phi(v_x\Phi(v_y)\Phi(v_z)$, allowing us to write

$$\boxed{\Phi(v_x)=\left(\frac{m}{2\pi kT}\right)^{1/2}e^{-mv_x^2/2kT}.} \tag{17-21}$$

A plot of $\Phi(v_x)$ against v_x gives the familiar bell-shaped curve, commonly called **Gaussian**, shown in Fig. 17-2. Several things may be noted: The most probable value of v_x is that of the maximum of the curve, or zero. At zero, Φ is $\sqrt{m/2\pi kT}$. Since the curve is symmetric about $v_x = 0$, positive and negative velocities are equally probable. The *average* velocity in the x-direction is clearly zero. If this were false and the average particle were moving, the entire box of fluid would have to be moving along the table (or floor). Although we have singled out $\Phi(v_x)$, the distributions $\Phi(v_y)$ and $\Phi(v_z)$ are identical in all respects. There are no special, different, or preferred directions in a simple fluid at equilibrium in the absence of an external field. Whatever direction we chose for the x-axis would yield the same result for Eq. 17-21 or Fig. 17-2.

The area under the curve of Fig. 17-2 between any two values of v_x is the probability that a particle in the system will have its x-component of velocity between the two limits. The total area under the curve must of course be unity, the probability of a certainty.

In Fig. 17-3, curves similar to the one in Fig. 17-2 are presented for three temperatures, $T_1 < T_2 < T_3$, to reveal how the distribution changes with temperature. At $v_x = \pm\sqrt{1.38kT/m}$, the value of $\Phi(v_x)$ has been damped to one-half its maximum. Therefore, at low temperatures, molecular velocities are clustered more heavily about zero. As the temperature is increased, fewer particles have low velocities and correspondingly more have high velocities.

This book always treats *velocity* as a directed or *vector* quantity. Since the system as a whole is not translating, the average velocity $\bar{v}$ is zero.

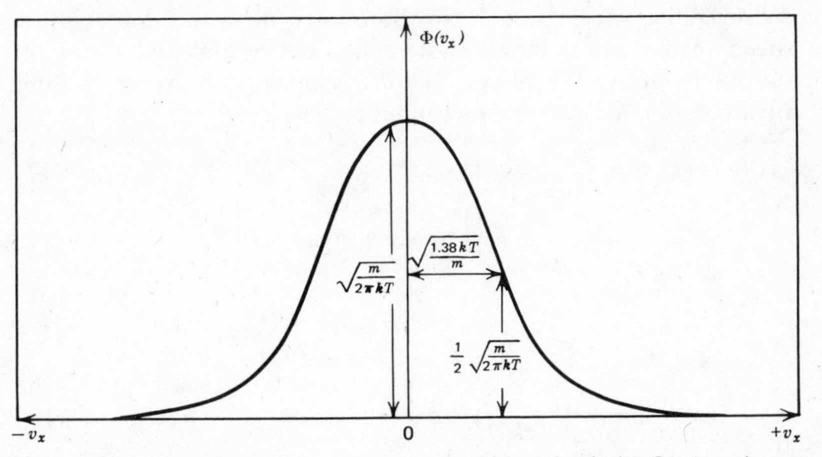

Fig. 17-2. Plot of distribution function for x-component of velocity $\Phi(v_x)$ against v_x, as given by Eq. 17-21.

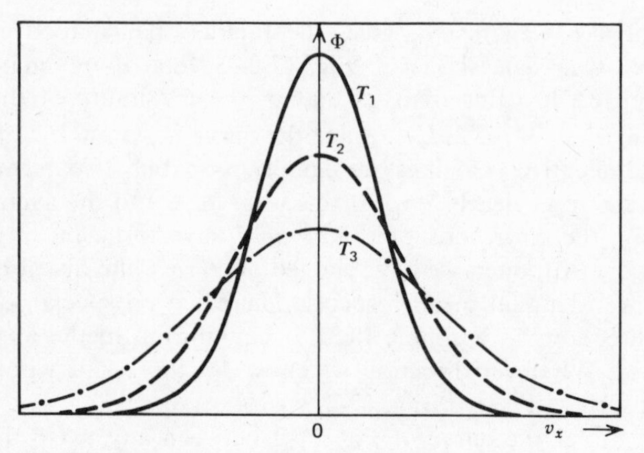

Fig. 17-3. Plot similar to Fig. 17-2, except for three different temperatures, $T_1, T_2 = 2T_1, T_3 = 4T_1$.

Equal contributions acting in opposite directions cancel to make the average of this vector quantity zero. Often we are less interested in the distribution of velocities, with their directions, than we are in the distribution of molecular speeds. In this book the word **speed** is always used for the *magnitude* of the velocity vector. The speed v is given by

$$v = |\mathbf{v}| = \sqrt{v_x^2 + v_y^2 + v_z^2} = \sqrt{v^2}\ . \tag{17-22}$$

The velocity vector may be characterized by its magnitude (the speed) and two angles (θ and χ), exactly as the momentum was pictured in Fig. 17-1*b*.

The last reduced distribution function we discuss is the **one-particle speed distribution** $\phi(v)$, defined as follows: $\phi(v)\,dv$ is the probability that the speed of a particle in the system lies between v and $v + dv$. It is normalized to unity. We obtain $\phi(v)\,dv$ from $\Phi_1(\mathbf{v})\,d\mathbf{v}$ by integrating in spherical coordinates over the unwanted angles:

$$\phi(v)\,dv = \int_{\text{angles}} \Phi_1(\mathbf{v})\,d\mathbf{v}, \tag{17-23}$$

$$= \int_{\text{angles}} \Phi_1(\mathbf{v})\,v^2\,dv\,\sin\theta\,d\theta\,d\chi, \tag{17-24}$$

$$= \left(\frac{m}{2\pi kT}\right)^{3/2} e^{-mv^2/2kT} v^2\,dv \int_0^{\pi} \sin\theta\,d\theta \int_0^{2\pi} d\chi, \tag{17-25}$$

$$= 4\pi\left(\frac{m}{2\pi kT}\right)^{3/2} e^{-mv^2/2kT} v^2\,dv, \tag{17-26}$$

$$\phi(v) = 4\pi\left(\frac{m}{2\pi kT}\right)^{3/2} v^2 e^{-mv^2/2kT}. \tag{17-27}$$

The form of the speed distribution $\phi(v)$ differs from that of the velocity distribution $\Phi_1(\mathbf{v})$ in its normalization constant and in the presence of the v^2 multiplying the exponential. Speeds, of course, are always positive; thus $\phi(v)$ is normalized to unity over the speed range zero to infinity. In Fig. 17-4, $\phi(v)$ is plotted against v. The curve is humped-shape because $\phi(v)$ is the product of two factors, the exponential (which drops off like the right-hand half of Fig. 17-2), and v^2 (which increases parabolically). For a while the increasing v^2 dominates and $\phi(v)$ increases. Then the decreasing Gaussian more than damps the v^2. The area under the curve between any two values of v is the probability that a particle in the system will have its speed somewhere between the two limits. The most probable value of each component of the *velocity* is zero, but the likelihood of zero *speed* is negligible. A zero speed demands that all three components of velocity be zero. Even though zero is the *most* probable value for each, it is *not very probable* for any one; thus the chance that all three components will be zero is negligible.

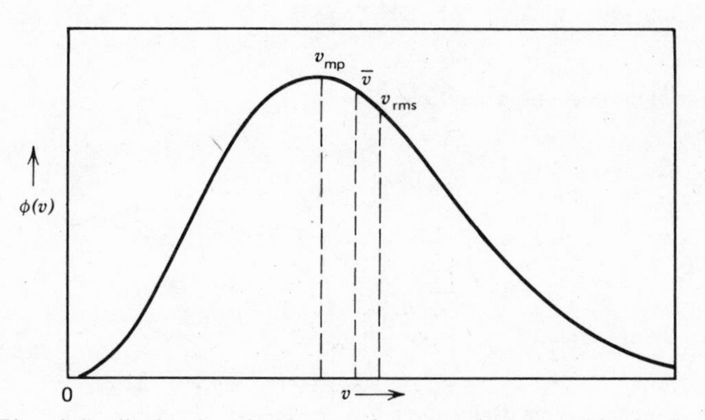

Fig. 17-4. Plot of distribution function for speed ϕ (v) against v, as given by Eq. 17-27.

AVERAGE SPEEDS AND ENERGIES

The **most probable speed** v_{mp} for a particle in the system is the value of v at the maximum of the speed distribution function $\phi(v)$. This maximum is found by putting the derivative of ϕ equal to zero:

$$\frac{d\phi}{dv}=\frac{d\phi}{d(v^2)}\frac{d(v^2)}{dv}=0; \qquad \frac{d\phi}{d(v^2)}=0, \tag{17-28}$$

$$4\pi\left(\frac{m}{2\pi kT}\right)^{3/2}\left(-\frac{m}{2kT}v^2e^{-mv^2/2kT}+e^{-mv^2/2kT}\right)=0, \tag{17-29}$$

$$-\frac{m}{2kT}v_{\mathrm{mp}}^2+1=0, \tag{17-30}$$

$$v_{\mathrm{mp}}^2=\frac{2kT}{m}=\frac{2RT}{M}, \tag{17-31}$$

$$\boxed{v_{\mathrm{mp}}=\sqrt{2}\sqrt{\frac{kT}{m}}=\sqrt{2}\sqrt{\frac{RT}{M}}\ .} \tag{17-32}$$

The symbol M designates the *molecular weight*, that is, the mass of Avogadro's number of particles:

$$M=N_0m \tag{17-33}$$

The **average speed of a particle** $\bar{v}$ is

$$\bar{v}=\int_0^\infty v\phi(v)\,dv, \tag{17-34}$$

$$=4\pi\left(\frac{m}{2\pi kT}\right)^{3/2}\int_0^\infty v^3e^{-mv^2/2kT}\,dv, \tag{17-35}$$

$$=4\pi\left(\frac{m}{2\pi kT}\right)^{3/2}\frac{(2kT)^2}{2m^2}, \tag{17-36}$$

$$\boxed{\bar{v}=\sqrt{\frac{8}{\pi}}\sqrt{\frac{kT}{m}}=\sqrt{\frac{8}{\pi}}\sqrt{\frac{RT}{M}}\ .} \tag{17-37}$$

The value of the integral is given in Eq. B-4 of Appendix B.

The **root mean square velocity** v_{rms} is of course the square root of $\overline{v^2}$:

$$\overline{v^2} = \int_0^\infty v^2 \phi(v)\, dv, \tag{17-38}$$

$$= 4\pi \left(\frac{m}{2\pi kT} \right)^{3/2} \int_0^\infty v^4 e^{-mv^2/2kT}\, dv, \tag{17-39}$$

$$= 4\pi \left(\frac{m}{2\pi kT} \right)^{3/2} \frac{3\pi^{1/2}}{8} \left(\frac{2kT}{m} \right)^{5/2}, \tag{17-40}$$

$$\overline{v^2} = \frac{3kT}{m} = \frac{3RT}{M}. \tag{17-41}$$

The value of the integral was obtained from Eq. B-5 in Appendix B. This is simply the classical derivation of the quantum result already obtained, Eq. 10-24:

$$\boxed{\bar{\epsilon}_{\text{trans}} = \tfrac{1}{2} m \overline{v^2} = \tfrac{3}{2} kT.} \tag{17-42}$$

Equations 17-41 and 17-43 can be remembered easily in the form of this important result. The average kinetic energy of a single particle in a gas or liquid is $\frac{3}{2}kT$; for a mole of particles it is $\frac{3}{2}RT$. The result is not only independent of density, but is also independent of the mass of the molecules. For example, in an equilibrium mixture of N_2, O_2, Ar, H_2O, CO_2, and large pollutant molecules, the average kinetic energy per molecule is the same, $\frac{3}{2}kT$, regardless of the kind of molecule. Finally, we have

$$\boxed{v_{\text{rms}} = \sqrt{\overline{v^2}} = \sqrt{3} \sqrt{\frac{kT}{m}} = \sqrt{3} \sqrt{\frac{RT}{M}}.} \tag{17-43}$$

We note that each of the three measures of typical molecular speeds in a fluid, v_{mp}, $\bar{v}$, and v_{rms}, is proportional to $\sqrt{kT/m}$ or $\sqrt{RT/M}$. Only the numerical coefficients are slightly different among them:

$$v_{\text{mp}} : \bar{v} : v_{\text{rms}} :: \sqrt{2} : \sqrt{\frac{8}{\pi}} : \sqrt{3} :: 1.41 : 1.59 : 1.73. \tag{17-44}$$

These three measures of typical speed appear in Fig. 17-4. In Fig. 17-5, the speed distribution is sketched for three different temperatures, $T_1, T_2 = 2T_1$, and $T_3 = 4T_1$. As T increases, the three averages increase as $\sqrt{T}$. The clustering of expected speeds about the most probable becomes less pronounced at higher temperatures because increased speeds become much more likely. Thus the temperature is a direct measure of average kinetic energy of a particle, and $\sqrt{T}$ is a measure of the average particle speed.

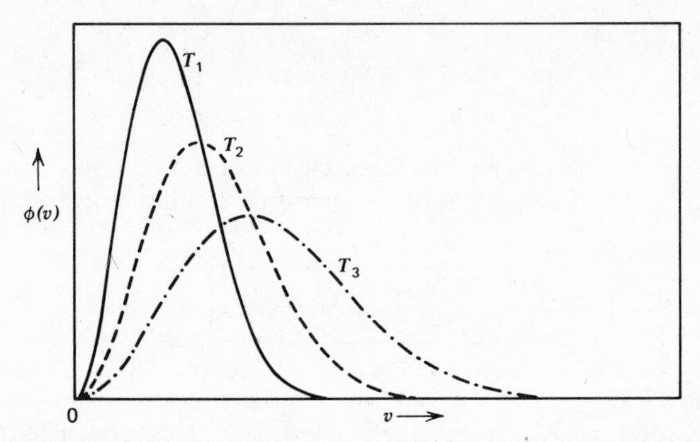

Fig. 17-5. Speed distribution plotted for three different temperatures, $T_1, T_2 = 2T_1, T_3 = 4T_1$.

By way of comparison with the average speeds just given, it is worth noting that the **speed of sound** propagated through a dilute gas is given by

$$v_{\text{sound}} = \sqrt{\gamma}\sqrt{\frac{kT}{m}} = \sqrt{\gamma}\sqrt{\frac{RT}{M}}\,, \tag{17-45}$$

where $\gamma = C_p / C_V = (C_V + nR)/C_V$. For monatomic gases $\gamma = \frac{5}{3}$; for diatomic gases $\gamma = \frac{7}{5}$. Thus the speeds of typical molecules in a fluid are only slightly faster than the speed of sound in the fluid at the same temperature.

We note the striking difference between, say, Fig. 17-4 for the speed distribution of individual particles and Fig. 5-3 for the energy of a macroscopic system in equilibrium with a heat bath. Although $\frac{3}{2}kT$ is the average particle kinetic energy, the kinetic energies of individual particles fluctuate greatly about $\frac{3}{2}kT$. However, in enormous numbers, these relatively unpredictable individuals exhibit highly predictable collective behavior. The kinetic energy contribution to the internal energy of almost any macroscopic fluid is exactly $\frac{3}{2}NkT$.

In conclusion, it is important to emphasize that nothing in this chapter is limited to dilute gases or even to gases. *Every result in this chapter pertains to the molecules in any simple fluid*, whether its density is that of a very dilute gas, a highly compressed liquid, or any substance in between.

PROBLEMS

17-1. Normalize with all particles in the system. Note: sometimes this average is calculated using only the particles moving from left to right, which gives twice the result. (Prove that.)

17-2. Verify that $\Phi_1(\mathbf{v})$ as given by Eq. 17-20 is properly normalized.

17-3. Prove that $\Phi(v_x)$ has decreased to half its maximum value at $v_x = \pm\sqrt{1.38kT/m}$.

17-4. In deriving Eq. 17-36, use was made of Eq. B-4. Prove Eq. B-2 and then Eq. B-4 from B-2.

17-5. In deriving Eq. 17-40, use was made of Eq. B-5. Prove Eq. B-3 and then Eq. B-5, starting with Eq. B-1.

17-6. In a mixture of H_2 and O_2, what is the ratio of $v_{\rm rms}(H_2)$ to $v_{\rm rms}(O_2)$? What is the same ratio of $\bar{v}$'s? Of $v_{\rm mp}$'s?

17-7. Calculate $\bar{v}$ for hydrogen and nitrogen gases at 25 C.

17-8. What fraction of members of the ensemble have particle i with *kinetic energy* between κ and $\kappa + d\kappa$? What kinetic energy is the most probable?

17-9. What interpretation should be given to the following velocity distribution, in which a, b, and c are constants:

$$\Phi_1(\mathbf{v}) = \left(\frac{m}{2\pi kT}\right)^{3/2} \exp\left\{-\frac{m}{2kT}\left[(v_x - a)^2 + (v_y - b)^2 + (v_z - c)^2\right]\right\}? \quad (17\text{-}46)$$

17-10. If an atom is radiating light of wavelength λ_0 and that atom is moving away from the observer at a rate of v_z cm/sec (where v_z is the z-component of its velocity), the observer will see the light at wavelength

$$\lambda = \lambda_0\left(1 + \frac{v_z}{c}\right), \quad (17\text{-}47)$$

because of the Doppler effect. The difference in speed between emitter and observer leads to an observed wavelength different from what would be seen if both had the same velocity. The factor c in Eq. 17-47 is the velocity of light. Suppose the atoms in a fluid at temperature T radiate at λ_0. An observer looking at the fluid from some distance away (far enough that all light rays reaching him are essentially parallel) will see the spectral line at λ_0 fuzzed, rather than sharp, because of the variety of Doppler shifts. If $I(\lambda)\,d\lambda$ is the intensity of illumination seen by the observer with wavelength between λ and $\lambda + d\lambda$, what is the predicted dependence of $I(\lambda)$ on λ in terms of I, where I is the total intensity the observer sees? Sketch a plot of $I(\lambda)$ versus λ. What fraction of the total energy has wavelength between 0.99 and 1.01 λ_0?

17-11. Two particular definite integrals, which cannot be evaluated in simpler terms but have been tabulated as functions of their upper (or lower) limit, are the so-called **error function of x**, abbreviated erf x, and the **complementary error function of x**, abbreviated erfc x, defined by

$$\operatorname{erf} x \equiv \frac{2}{\sqrt{\pi}} \int_0^x e^{-z^2}\,dz; \qquad \operatorname{erfc} x \equiv \frac{2}{\sqrt{\pi}} \int_x^\infty e^{-z^2}\,dz. \tag{17-48}$$

What is the value of erf y, where $y \ll 1$? What is the value of erf y where $y \gg 1$? What is the value of erfc x in terms of the value of erf x? What fraction of molecules in a fluid have x-component of velocity between $\pm\sqrt{kT/m}$?

17-12. Suppose the system contained only three particles and was in equilibrium with a heat bath at temperature T; how would that affect the form of $\Phi_1(\mathbf{v})$? If instead of a heat bath, this system of three particles were isolated with fixed energy, how would that affect the form of $\Phi_1(\mathbf{v})$ or the shape of Fig. 17-2?

17-13. What is the speed in kilometers per second at which $\Phi_1(v)$ takes just half its maximum value for nitrogen molecules at 25 C? Compare with v_{mp}. How can the most probable speed be greater than the speed that damps $\Phi_1(v)$ to half its maximum value?

18 THE CLASSICAL GAS*

THE PRESSURE IN AN IDEAL GAS

In a dilute gas, the molecules are so far apart that they fly freely almost all the time, very rarely colliding with a wall or with another molecule. Here we must visualize the molecular behavior that gives rise to the macroscopic pressure. Pressure in a region is defined as the force the gas would exert on a test surface of unit area located in that region. The force is normal to the test surface, and since the force is the

*Certain aspects of this classical mechanical study are frequently referred to as the *kinetic theory of gases,* but different people give the phrase different meanings. Because of this lack of agreement, the phrase "kinetic theory" is not used in this book.

time rate of change of momentum, the pressure can be written

$$P = \frac{\text{force normal to surface}}{\text{area of surface}}, \tag{18-1}$$

$$\bar{P} = \left[\frac{\text{change of } \mathbf{p} \text{ normal to surface}}{(\text{area of surface})(\text{time for that } \Delta\mathbf{p})} \right]_{\text{avg}}. \tag{18-2}$$

Let the area A be the top side of the test surface located perpendicular to the z-axis, as in Fig. 18-1. Every particle approaching A from the $+z$ side will rebound after striking A. The p_x and p_y components of momentum of those particles will be unchanged if the collisions are elastic (which they will be, on average), but the p_z component will be changed in sign. Thus the total change in momentum of the particle due to the collision is twice the magnitude of the value of p_z after the collision (p_z must be negative before the collision, positive afterwards, for the particle to strike A). Thus we have

$$\bar{P} = \left[\frac{\Delta\mathbf{p}}{At} \right]_{\text{avg}} = \left[\frac{2p_z}{At} \right]_{\text{avg}}. \tag{18-3}$$

We average Eq. 18-3, using $f_1(\mathbf{r},\mathbf{p})$ as given by Eq. 17-12. The components of momentum p_x and p_y range from $-\infty$ to $+\infty$, whereas p_z will range from $-\infty$ to 0:

$$\bar{P} = \frac{2}{At} \int_{-\infty}^{\infty} dp_x \int_{-\infty}^{\infty} dp_y \int_{-\infty}^{0} dp_z \int_{\mathrm{V}} d\mathbf{r} |p_z| f_1(\mathbf{r},\mathbf{p}). \tag{18-4}$$

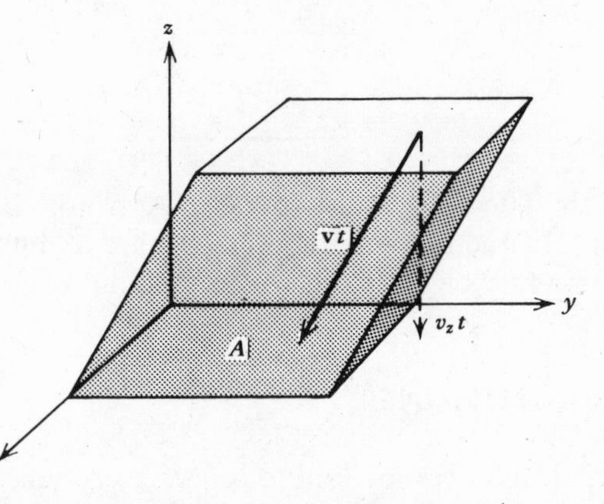

Fig. 18-1. Parallelepiped of impact for a particular edge, $\mathbf{v}t = \mathbf{p}t/m$.

Here, v is the volume within which particles must be at time 0 if they are to strike A during the time t. This is the volume of the parallelepiped having A as its base and its edges given by $\mathbf{v}t = \mathbf{p}t/m$ (Fig. 18-1). Since there is a different parallelepiped for each velocity, the volume of the parallelepiped is a function of velocity. This volume is the area A of the base times its height $|p_z|t/m$. Thus the volume v within which a particle must be if it is to strike A from above during the time t is

$$\mathrm{v} = \frac{A|p_z|t}{m}. \tag{18-5}$$

Since the integrand of Eq. 18-4 is independent of $\mathbf{r}$, the integral over $d\mathbf{r}$ simply yields v. The average pressure therefore is

$$\bar{P} = \frac{2}{At}\int_{-\infty}^{\infty} dp_x \int_{-\infty}^{\infty} dp_y \int_{-\infty}^{0} dp_z |p_z| \mathrm{v} f_1(\mathbf{r},\mathbf{p}), \tag{18-6}$$

$$= \frac{2}{At}\int_{-\infty}^{\infty} dp_x \int_{-\infty}^{\infty} dp_y \int_{-\infty}^{0} dp_z |p_z| \left(\frac{A|p_z|t}{m}\right) \frac{N}{V}(2\pi mkT)^{-3/2} e^{-p^2/2mkT}, \tag{18-7}$$

$$= \frac{2N}{Vm}(2\pi mkT)^{-3/2} \int_{-\infty}^{\infty} dp_x e^{-p_x^2/2mkT} \int_{-\infty}^{\infty} dp_y e^{-p_y^2/2mkT} \int_{-\infty}^{0} dp_z p_z^2 e^{-p_z^2/2mkT}, \tag{18-8}$$

$$= \frac{2N}{Vm}(2\pi mkT)^{-3/2}(2\pi mkT)^{1/2}(2\pi mkT)^{1/2}\frac{\pi^{1/2}}{4}(2mkT)^{3/2}, \tag{18-9}$$

$$\boxed{P = \frac{NkT}{V}.} \tag{18-10}$$

The values of the integrals can be found in Appendix B. This, of course, is the same result obtained by quantum mechanics, and it is the experimental value for the pressure of all real gases in the limit of vanishing density.

MOLECULAR COLLISIONS

Suppose two molecules can be said to *collide* every time their centers come within a distance σ of each other. Different values of σ might prove useful for different processes, but for a given process a choice of σ leads to the so-called **collision cross section** $s = \pi\sigma^2$. The cross section is the effective area that one molecule presents to another as they move by each other during the process of interest. If the particles are hard spheres, the definition of a collision is unambiguous (see Fig. 18-2). The diameter of each molecule is σ; the intermolecular potential energy curve of Fig. 10-3 would show an infinite potential for $|\mathbf{r}_2 - \mathbf{r}_1|$ (the distance between the centers of particles 1 and 2) less than σ and a zero potential for $|\mathbf{r}_2 - \mathbf{r}_1|$ greater than σ. With more realistic intermolecular potentials, such as those shown in Fig. 10-3, it may not be certain whether a collision has occurred, since the long-distance tail of the potential approaches the value zero gradually. This difficulty is more common in nonequilibrium processes; we shall not worry about it here.

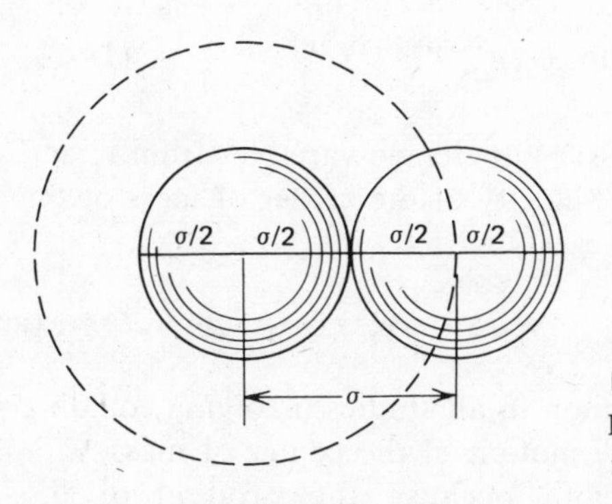

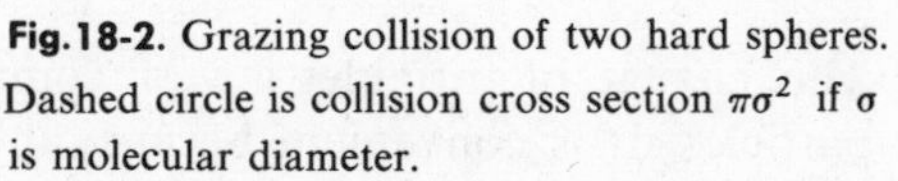

Fig. 18-2. Grazing collision of two hard spheres. Dashed circle is collision cross section $\pi\sigma^2$ if σ is molecular diameter.

Suppose a person is riding on particle 1 as it moved around. Concerned about the frequency of collisions his molecule would experience, he looks about at the nearby molecules to see how they are moving. The velocity that he sees for, say, particle 2 is the **relative velocity** $\mathbf{v}_{21}$ of particle 2:

$$\mathbf{v}_{21} \equiv \mathbf{v}_2 - \mathbf{v}_1, \tag{18-11}$$

which is the time rate of change of $\mathbf{r}_{21} \equiv \mathbf{r}_2 - \mathbf{r}_1$. The difference between $\mathbf{v}_2$ and $\mathbf{v}_{21}$ is that the vector $\mathbf{v}_2$ has an origin that is at rest with respect to the laboratory, whereas $\mathbf{v}_{21}$ has as its origin the velocity of particle 1, which of course varies. Our rider would see particle 2 sweep out a collision volume $sv_{21}t$ during the time t, where v_{21} is the magnitude of $\mathbf{v}_{21}$ (i.e., $v_{21} = |\mathbf{v}_{21}| = |\mathbf{v}_2 - \mathbf{v}_1|$). This is represented in Fig. 18-3. It is the *magnitude* alone of $\mathbf{v}_{21}$ that determines the collision volume; the angles merely determine the orientation of that volume in space.

Fig. 18-3. Volume an observer sitting on particle 1 sees particle 2 sweep out during time t, $sv_{21}t$.

We now calculate the ensemble average of the volume our observer sees swept out by particle 2 as he rides on particle 1. This requires finding the average relative velocity between any two molecules. Since their velocities are uncorrelated, $\overline{v_{21}}$ is

$$\overline{v_{21}} = \int d\mathbf{v}_1\, d\mathbf{v}_2\, v_{21} \Phi_1(\mathbf{v}_1) \Phi_1(\mathbf{v}_2), \tag{18-12}$$

$$\overline{v_{21}} = \left(\frac{m}{2\pi kT}\right)^3 \int d\mathbf{v}_1\, d\mathbf{v}_2\, v_{21}\, e^{-m(v_1^2+v_2^2)/2kT}. \tag{18-13}$$

To perform the integration, it is necessary to change variables from $\mathbf{v}_1$ and $\mathbf{v}_2$ to the relative velocity $\mathbf{v}_{21}$ and the velocity of the center of mass of the molecules $\mathbf{V}_{21}$:

$$\mathbf{V}_{21} \equiv \tfrac{1}{2}(\mathbf{v}_2 + \mathbf{v}_1); \qquad \mathbf{v}_{21} \equiv \mathbf{v}_2 - \mathbf{v}_1. \tag{18-14}$$

This change of variables is very common in all studies involving colliding particles. It is convenient because the motion of the center of mass $\mathbf{V}_{21}$ is undisturbed during the collision and because the product of the differentials is preserved, as is proved in Prob. 18-15.

$$d\mathbf{v}_1\, d\mathbf{v}_2 = d\mathbf{V}_{21}\, d\mathbf{v}_{21}. \tag{18-15}$$

We note also that

$$v_1^2 + v_2^2 = 2V_{21}^2 + \tfrac{1}{2}v_{21}^2, \tag{18-16}$$

which can be checked immediately by the reader. Changed to center of mass and relative velocities, Eq. 18-13 becomes

$$\overline{v_{21}} = \left(\frac{m}{2\pi kT}\right)^3 \int d\mathbf{V}_{21}\, d\mathbf{v}_{21} v_{21} e^{-m[2V_{21}^2 + (1/2)v_{21}^2]/2kT}. \tag{18-17}$$

The integration over $\mathbf{V}_{21}$ can be done immediately to yield

$$\overline{v_{21}} = \left(\frac{m}{2\pi kT}\right)^3 \left(\frac{\pi kT}{m}\right)^{3/2} \int d\mathbf{v}_{21}\, v_{21}\, e^{-mv_{21}^2/4kT}. \tag{18-18}$$

To perform the integration over $d\mathbf{v}_{21}$, this three-dimensional element of volume in velocity space is put into spherical coordinates $d\mathbf{v}_{21} = v_{21}^2\, dv_{21} \sin\theta\, d\theta\, d\chi$. Since there is no dependence in the integrand on the angles, their integration gives 4π. The remaining integral is

$$\overline{v_{21}} = 4\pi\left(\frac{m}{2\pi kT}\right)^3\left(\frac{\pi kT}{m}\right)^{3/2}\int_0^\infty dv_{21}\, v_{21}^3\, e^{-mv_{21}^2/4kT}, \tag{18-19}$$

$$= 4\pi\left(\frac{m}{2\pi kT}\right)^3\left(\frac{\pi kT}{m}\right)^{3/2}\frac{1}{2}\left(\frac{4kT}{m}\right)^2, \tag{18-20}$$

$$= 4\left(\frac{kT}{\pi m}\right)^{1/2} = \sqrt{2}\sqrt{\frac{8}{\pi}}\sqrt{\frac{kT}{m}}, \tag{18-21}$$

$$\boxed{\overline{v_{21}} = \sqrt{2}\ \bar{v}.} \tag{18-22}$$

Interestingly, the average speed of the particles in the system relative to the laboratory is $\bar{v}$, whereas the average speed relative to an observer riding on one of the particles would be $\sqrt{2}\ \bar{v}$. This relative speed is the same as that of two molecules that began at the same time, moving outward at right angles to each other at speeds $\bar{v}$.*

We can now find the **average frequency of collisions** Z_1 for a single particle in the system. A rider on that particle sees all $N-1$ other particles in the system sweeping out collision volumes of $s\,\overline{v_{21}}\,t$, on average, during the time t (we commonly replace $N-1$ by N, since 1 means nothing to a number like 10^{20}). The probability that a particular particle lies in any small volume $Ns\,\overline{v_{21}}\,t$ is simply the ratio of that volume to the total volume V. Thus the probability that the particle of interest collides with some other particle during t is $Ns\,\overline{v_{21}}\,t/V$. The average collision frequency is this divided by t:

$$Z_1 = \frac{Ns\,\overline{v_{21}}}{V} = \rho s\,\overline{v_{21}} = \sqrt{2}\ \rho s\bar{v} = 4\rho s\left(\frac{kT}{\pi m}\right)^{1/2} = 4\rho s\left(\frac{RT}{\pi M}\right)^{1/2}. \tag{18-23}$$

*Often in elementary texts such intuitive statements are used as "proofs"of the results. For example, all the molecules are sometimes divided into three groups, each moving parallel to one of the three coordinate axes. It is safe to say about most such elementary "proofs" of results like Eq. 18-10, 18-22, 18-25, and 18-31, that they are pictorial plausibility arguments; they give the right answers because the answers were known in advance; as "proofs," however, they are fraudulent.

We demonstrate in Prob. 18-1 that for a gas like nitrogen at 1 atm pressure and 25 C, if s is about 60 Å^2, Z_1 is about 10^{10} collisions per second.

Now, if each individual molecule experiences on average Z_1 collisions per second, NZ_1 is the total number of collisions experienced by the N molecules. Since each collision involves two molecules, the **total number of collisions** Z is half this value:

$$Z = \tfrac{1}{2} N Z_1. \tag{18-24}$$

Sometimes one perfers to employ the **total number of collisions per unit volume** z, which is just Z/V:

$$z = \frac{Z}{V} = \frac{1}{2}\rho Z_1 = \frac{1}{\sqrt{2}} s\rho^2 \bar{v} = 2s\rho^2\left(\frac{RT}{\pi M}\right)^{1/2}. \tag{18-25}$$

It is shown in Prob. 18-2 that in each cubic centimeter of the nitrogen gas mentioned previously there are about 10^{29} collisions per second.

The average or **mean free path** L between collisions for a given particle is found by noting that a molecule traveling an average distance of $\bar{v}$ per second and colliding an average of Z_1 times per second goes an average distance of

$$L = \frac{\bar{v}}{Z_1} = \frac{1}{\sqrt{2}\, s\rho} = \frac{kT}{\sqrt{2}\, sP} \tag{18-26}$$

between collisions. Thus the mean free path is inversely proportional to the density and (if s is truly independent of T) at constant ρ is independent of temperature. At constant T it is inversely proportional to P. The mean free path of Eq. 18-26 is a difficult type of average to define in terms of the ensemble. It does, however, give an idea of what is going on in the gas. It is indicated in Prob. 18-3 that for the above-mentioned nitrogen gas L is about 500 Å, some 100 times as great as the radius of a molecule. This is consistent with our view of molecules in dilute gases as spending most of their time flying freely between collisions.

Our last average quantity describing molecular collisions is the rate at which particles will strike an area A, which might be part of the wall. This rate can be measured experimentally by poking a small hole in the wall of the vessel and measuring the outflow of gas. The hole made is tiny, of diameter less than the mean free path, preventing the outward rush of gas from disturbing the equilibrium velocity distribution on which the whole theory is based. Such a process is called **molecular effusion**. It is analogous to the study of blackbody radiation in Chap. 15. The problem of finding the **number of particles striking unit area per unit time** Γ (or the rate of effusion) is an average calculated exactly like that of the pressure in Eqs. 18-4 to 18-10. The only difference is that P is the average, Eq. 18-3, of $2p_z/At$ for the collisions, whereas

$$\Gamma = \left[\frac{1}{At}\right]_{\text{avg}}, \tag{18-27}$$

where the unity simply means we are counting collisions one at a time. Thus Γ is given by Eq. 18-6 with the replacement of $2p_z$ by 1:

$$\Gamma = \frac{1}{At}\int_{-\infty}^{\infty} dp_x \int_{-\infty}^{\infty} dp_y \int_{-\infty}^{0} dp_z\, v f_1(\mathbf{r},\mathbf{p}), \tag{18-28}$$

$$= \frac{N}{mV}(2\pi mkT)^{-3/2}\int_{-\infty}^{\infty} dp_x \int_{-\infty}^{\infty} dp_y \int_{-\infty}^{0} dp_z |p_z| e^{-(p_x^2+p_y^2+p_z^2)/2mkT}, \tag{18-29}$$

$$= \frac{N}{mV}(2\pi mkT)^{-3/2}(2\pi mkT)^{1/2}(2\pi mkT)^{1/2}(mkT), \tag{18-30}$$

$$\Gamma = \frac{N}{V}\sqrt{\frac{kT}{2\pi m}} = \rho\sqrt{\frac{RT}{2\pi M}} = \frac{1}{4}\rho\bar{v}. \tag{18-31}$$

The rate of effusion through the hole is the same as if the entire gas were moving bodily through it with speed $\frac{1}{4}\bar{v}$. The same result was obtained for radiation in Chap. 15.

Obviously, if two chambers containing gases are separated by a small hole, gas will simultaneously effuse in both directions. If the hole dimensions are small compared with L, it is very unlikely that molecules will strike one another in the immediate vicinity of the hole; thus Eq. 18-31 works in both directions at the same time. Eventually, if the gas on one side or the other is neither replenished nor pumped off, equilibrium will be reached. Interestingly enough, even for holes too large, where the equilibrium conditions are clearly destroyed, Eq. 18-31 stays surprisingly accurate because of lucky cancellation of errors. A number of elementary laboratory experiments designed to illustrate Eq. 18-31 have far too large a hole for the pressure of the gas, yet results agree with the "theory."

CLASSICAL EQUIPARTITION OF ENERGY

We have seen in Eq. 17-42 how the three components of velocity, which contribute $\frac{1}{2}mv_x^2$, $\frac{1}{2}mv_y^2$, and $\frac{1}{2}mv_z^2$ to the energy of the molecule, each contribute $\frac{1}{2}kT$ to its average energy. In classical mechanics, kinetic energy is not the only kind of energy whose amount is proportional to the square of a phase space variable. Rotational energy about each axis is given by $\frac{1}{2}I\omega^2$, where ω is the angular velocity and I the moment of inertia. Vibrational energy for each normal mode is given by $p^2/2\mu + Kq^2$, where p is a generalized momentum for the vibration, μ a generalized or reduced mass, q a generalized coordinate that measures the extent of the vibration, and K the force constant for the harmonic vibrational potential. The $p^2/2\mu$ represents the kinetic and Kq^2 the potential vibrational energy. Thus the total energy of a classical molecule of n atoms contains three terms $\frac{1}{2}mv_i^2$ for the center of mass, two (for linear molecules, three for nonlinear) terms $\frac{1}{2}I_i\omega_i^2$ for rotation, and $3n-5$ (for linear molecules, $3n-6$ for nonlinear) terms $p_i^2/2\mu_i + K_iq_i^2$ for vibration.

For Boltzmann statistics, **the classical ensemble average of each such contribution to the energy is** $\frac{1}{2}kT$. The proof is essentially that of Eq. 17-42. The amount of energy in each contribution is independent of the amounts in the other contributions, the total energy E being the sum of the contributions ϵ, $e^{-\beta E}$ factoring into separate terms for each ϵ. Thus the average value of any contribution of the form $a\eta^2$ is

$$\bar{\epsilon} = \frac{\int d\eta\, a\eta^2 e^{-\beta a\eta^2}}{\int d\eta\, e^{-\beta a\eta^2}} = -\frac{\partial}{\partial\beta}\ln\int d\eta\, e^{-\beta a\eta^2}. \tag{18-32}$$

We must know the limits on the integral. If η refers to a translational momentum or velocity component, the limits are of course $-\infty$ to $+\infty$. The same holds for angular velocity. For vibrations, however, the limits on the p's and q's are less certain. However, it is enough to note that the integrand becomes vanishingly small for $p_i^2/2\mu_i$ or $K_iq_i^2$ much greater than kT, owing to the damping of the exponential. Therefore, whatever the actual limits on the p's and q's might be, they too may be treated as $-\infty$ to $+\infty$. Thus the value of the integral can be obtained from Eq. B-1, and Eq. 18-32 becomes simply

$$\bar{\epsilon} = \overline{a\eta^2} = -\frac{\partial}{\partial\beta}\ln\left(\frac{\pi}{\beta a}\right)^{1/2} = \frac{1}{2\beta} = \frac{1}{2}kT. \tag{18-33}$$

This conclusion is called the **equipartition of energy**.

SUMMARY

The classical mechanical average for the energy of a molecule of n atoms is $\frac{1}{2}kT$ for each of the three translational components and for each of the two (or three if nonlinear) rotational components. It is kT for each of the $3n-5$ (or $3n-6$ if nonlinear) normal modes of vibration. The average intermolecular potential energy in dense gases and liquids is a much more complicated problem, which is treated briefly in Chap. 20. The energy relative to external fields is an easier matter, and for gravity it is treated in Chap. 19. Molecules, which of course are really described by quantum mechanics and not classical, will reach their classical average energies only when T is so high that the separation of adjacent molecular quantum states is small compared with kT. This is called the **classical limit** of the quantum problem. At lower temperatures, those below the characteristic temperature Θ for the energy contribution being considered (see Chap. 11), the average energy will be below the classical value. The contribution is frozen out due to its wide level spacings compared with kT.

PROBLEMS

18-1. What is the average frequency of collisions of a nitrogen molecule in a vessel containing N_2 at 1.00 atm pressure and 20 C? Let $s = 60$ Å^2.

18-2. How many collisions will occur between the N_2 molecules of Prob. 18-1 in each cubic centimeter of the gas?

18-3. What is the mean free path for the N_2 molecules of Prob. 18-1? How many times the molecular radius is the mean free path?

18-4. What molar specific heat at constant volume would you expect under classical conditions for the following gases: (*a*) Ne, (*b*) O_2, (*c*) H_2O (nonlinear), (*d*) CO_2 (linear), (*e*) $CHCl_3$?

18-5. Compare the value of the average kinetic energy of a molecule at 25° C with the energies of Figs. 10-3 and 11-2. What does your result imply about the strengths of the bonds represented by those figures?

18-6. An excellent laboratory vacuum is 10^{-8} torr. How many gas molecules are still contained in 1 mm^3 (volume the size of a pinhead) at 300 K? What is the mean free path if the collision cross section is 30 Å^2?

18-7. Two chambers are separated by a membrane, pierced with small holes. One chamber contains H_2, the other O_2. Both gases are at the same temperature. The pressure in the oxygen chamber is kept at twice that in the other. What happens?

18-8. How many molecules of N_2 effuse each minute through a square hole 0.01 mm on a side at 0.1 atm and 25 C?

18-9. Note the similarity between Eqs. 18-23 and 18-31 and discuss the reason for it.

18-10. A **Knudsen cell** is a small chamber containing a material whose vapor pressure is desired as a function of temperature. The chamber, pierced by a small hole of measured dimensions, is weighed before and after a period during which it is thermostated in a vacuum. The effusion equation is presumed to hold. From the weight loss, the vapor pressure is calculated. The following is a typical teaching laboratory experiment. A Knudsen cell containing benzoic acid ($M = 122$ g/mole) is weighed and then thermostated at 70 C in an evacuated chamber for 60 min. The hole in the cell is circular, of 0.60 mm diameter. What is the sublimation pressure obtained if the weight loss from the cell is 56.7 mg? What is the ratio of hole diameter to mean free path in this experiment? Estimate the collision diameter of benzoic acid as 10 Å.

18-11. Prove that the number of collisions of gas molecules per unit area of wall per unit time in which the angle between the incoming direction of the particle and a normal to the surface is between θ and $\theta + d\theta$ is

$$\frac{N}{V}\left(\frac{2kT}{\pi m}\right)^{1/2} \sin\theta\cos\theta\, d\theta.$$

18-12. Find the number of collisions of gas molecules per unit area of wall per second in which the speed is between v and $v + dv$.

18-13. Calculate the root-mean-square velocity of the gas molecules effusing through a small hole. How does their average kinetic energy compare with that of the particles that do not effuse out. Why is this?

18-14. Prove the so-called **Clausius formula**, namely, that the probability a gas molecule will suffer *no* collision along a path of length equal to or less than l is $e^{-l/L}$, where L is the mean free path.

18-15. Consider some arbitrary plane in the gas. Prove that on the average the molecules on reaching that plane have had their last collision at a distance $2L/3$ from the plane.

18-16. Prove the validity of Eq. 18-15 for the x-components of the vectors. The same proof must of course hold for the y- and z-components. Refer to Eqs. 2-57 to 2-75.

18-17. In an ideal gas mixture, molecules of type A and type B have average speeds $\bar{v}_A$ and $\bar{v}_B$ given by Eq. 17-37, with m_A and m_B replacing m. Calculate the average relative speed $\bar{v}_{AB}$ between A- and B-type molecules.

18-18. In an ideal gas mixture with densities of species A and B equal to ρ_A and ρ_B, what is the total number of collisions per unit volume per second between A–A, B–B, and A–B pairs? The collision cross sections are s_{AA}, s_{BB}, and s_{AB}, respectively.

18-19. In a typical molecular beam experiment, atoms effuse from a furnace through a thin slit and are collimated to eliminate those that are not going in almost exactly the same direction, at right angles to the slit. What is the average speed of the molecules in the beam as function of atomic mass and temperature of the furnace? What is the value of this speed for sodium atoms in an oven at 740 K?

18-20. Some molecular beam experiments go beyond the setup of Prob. 18-19 to pass the beam through a velocity selector, which removes all particles whose velocity is outside the range $\bar{v} \pm \Delta v$. How large a fraction of $\bar{v}$ must Δv be made for the velocity selector not to reduce the intensity of the beam by more than an order of magnitude (i.e., a factor of 10)? By $\bar{v}$ here we mean the average speed of the effusing particles, as in Prob. 18-19.

19 GRAVITATIONAL FIELDS

To consider the effects of a gravitational field on a fluid, let us return to Eq. 17-7 for the one-particle distribution function,

$$f_1(\mathbf{r},\mathbf{p}) = Ae^{-\beta\epsilon(\mathbf{r},\mathbf{p})}, \tag{19-1}$$

where $\epsilon(\mathbf{r},\mathbf{p})$ is the average energy of a single particle in the fluid at the point $\mathbf{r}$, having the momentum $\mathbf{p}$. We will restrict ourselves to cases in which the changing gravitational potential does not change the density so much that the average molecular environment of the molecules being considered is modified appreciably. In this case, the average intermolecular potential energy can be treated as independent of r, and the only energies that are included in $\epsilon(r,p)$ are the kinetic and the gravitational. The **gravitational potential** $\psi(\mathbf{r})$ is defined as follows: The work required to move mass m_i from $\mathbf{r}_1$ to $\mathbf{r}_2$ is $m_i[\psi(\mathbf{r}_2) - \psi(\mathbf{r}_1)]$. In the fluid we can put

$$\epsilon_i(\mathbf{r},\mathbf{p}) = \frac{p_i^2}{2m_i} + m_i^*\psi(\mathbf{r}), \tag{19-2}$$

where m_i^* is the mass of the particle of type i less the weight of the fluid it displaces. For dilute gases, m_i^* can simply be relaced by m_i. For dense gases and liquids one must use

$$m_i^* = m_i - v_i d, \tag{19-3}$$

where v_i is the partial molecular volume of substance i and d the average mass density of the fluid at the point $\mathbf{r}$.

We can find the particle density $\rho(\mathbf{r})$ using Eq. 17-13, by integrating over the momenta:

$$\rho(\mathbf{r}) = Ae^{-m^*\psi(\mathbf{r})/kT}. \quad \text{(normalized to } N\text{)} \tag{19-4}$$

Often one is interested simply in the ratio of the densities at two different points, in which case the normalization constant need not be calculated:

$$\frac{\rho(\mathbf{r}_2)}{\rho(\mathbf{r}_1)} = e^{-m^*[\psi(\mathbf{r}_2)-\psi(\mathbf{r}_1)]/kT}. \tag{19-5}$$

For dilute gases, the pressure is proportional to the density:

$$P = \rho kT, \qquad \text{(ideal gas)} \tag{19-6}$$

as was derived in Eq. 18-10. This result can be used in the presence of a gravitational field so long as the volume v used in its derivation has dimensions small compared with the distance over which ρ changes appreciably because of gravity. In such a case,

$$P(\mathbf{r}) = \rho(\mathbf{r})kT = \alpha e^{-m\psi(\mathbf{r})/kT}. \qquad \text{(ideal gas)} \tag{19-7}$$

If the gravitational potential arises only from the earth, $\psi(\mathbf{r})$ at a height h above a reference elevation is

$$\psi(\mathbf{r}) = gh, \tag{19-8}$$

for heights small enough that g can be taken as constant. The acceleration of gravity g has the value 980 cm/sec^2 at sea level. Use of this with Eqs. 19-5 and 19-7 yields

$$\frac{P(h_2)}{P(h_1)} = \frac{\rho(h_2)}{\rho(h_1)} = e^{-mg(h_2-h_1)/kT} = e^{-Mg(h_2-h_1)/RT}. \tag{19-9}$$

This is the ratio of pressures or densities at two different heights in the earth's gravitational field for an ideal gas of molecular weight M, each molecule of which weighs m. The system must be at equilibrium, thus T must be the same at both heights. The density and pressure decrease the higher one goes. The greater the molecular weight of the gas, the more abrupt is its dropoff in density. Various forms of Eq. 19-9 are called the **barometric equation.** It can be derived from purely thermodynamic arguments, which ignore the particulate nature of the gas.

Another use of Eq. 19-5 is in understanding the sedimentation equilibrium of suspensions of colloidal particles in liquids. Colloidal particles are large by molecular standards, from about 10 to 10,000 Å. If they are suspended in a liquid, the constant bombardment by molecules of the liquid causes them to undergo Brownian motion and keeps them from settling out under the influence of gravity. The ratio of densities at two different heights is simply

$$\frac{\rho(h_2)}{\rho(h_1)} = e^{-m^*g(h_2-h_1)/kT} = e^{-M^*g(h_2-h_1)/RT}. \tag{19-10}$$

Of course, complications can enter these equations if the colloidal suspension is so concentrated that the particles interact with each other; the amount of interaction can then change with height. As any property of a colloidal suspension is plotted against ρ, the value extrapolated back to $\rho=0$ is always observed to be "correct," as given by this simple theory.

Many studies of colloidal suspensions have been made using Eq. 19-10, but a major drawback is the relative smallness of the earth's gravitational field. During the period 1923 to 1925, T. Svedberg developed the *ultracentrifuge,* which can increase the gravitational field to as much as several hundred thousand g.* As is known from elementary physics, if a particle of mass m^* is spun ν revolutions per second at a distance x from the axis of revolution, the centrifugal energy is

$$E_{\text{cent}} = -\frac{1}{2}m^*\nu^2x^2. \tag{19-11}$$

This can be verified quickly by noting that it yields a centrifugal force

$$F_{\text{cent}} = -\frac{\partial E_{\text{cent}}}{\partial x} = m^*\nu^2x, \tag{19-12}$$

which is probably more familiar to the reader. In this case, Eq. 19-5 becomes

$$\rho(x_2) = \rho(x_1)e^{-m^*\nu^2(x_1^2-x_2^2)/2kT} \tag{19-13}$$

or

$$\ln\rho(x_2) = -\frac{m^*\nu^2(x_1^2-x_2^2)}{2kT} + \ln\rho(x_1). \tag{19-14}$$

Thus one can find m^* by equating to $m^*\nu^2/2kT$ the slope of a plot of $\ln\rho(x)$ against x^2. The ultracentrifuge permits fairly accurate determination of the molecular weight of even reasonably small molecules by this method.

*This remarkable instrument and its application to a variety of problems are discussed by T. Svedberg and K. O. Pedersen, *The Ultracentrifuge* (Oxford University Press, London, 1940).

PROBLEMS

19-1. Suppose you are going up in an elevator and your ears pop at the thirtieth floor, about 100 m above the ground. What pressure change caused your ears to pop? Note: for small x, $e^{-x} \approx 1 - x$. The average molecular weight of air is about 29. The temperature could be 300 K.

19-2. Consider that at the earth's surface, air is composed mostly of oxygen and nitrogen with average molecular weight of 29. About one part in a thousand, however, is hydrogen with molecular weight of 2. If the air is at about -10 C to all heights, at what height does the concentration of hydrogen equal the concentration of oxygen and nitrogen put together? At 800 km, what is the ratio of the concentration of hydrogen to that of the other species? Assume that Eq. 19-9 is still valid.

19-3. Below what height will be found just half the gas molecules in the earth's gravitational field? Let T be constant at 0 C and g be constant at 980 cm/sec^2. The average molecular weight of air is 29.

19-4. What is the mean free path of air molecules (assumed to have an average molecular weight of 29 and a collision cross section of 60 Å^2) at heights of 10 and 100 miles above the earth's surface. Let the temperature be -10 C at all heights, and let g be constant at 980 cm/sec^2.

19-5. Find the density of an ideal gas at some arbitrary point in a cylinder of radius R and length l. The cylinder contains N molecules, each weighing m grams, and rotates about its axis ν times per second. Neglect the external gravitational field.

19-6. Experiments with colloids based on equations as simple as 19-13 provided some of the first *direct* evidence for the existence of molecules, one of the earliest tests of the statistical mechanics of macroscopic substances, and an accurate determination of Avogadro's number. J. Perrin and coworkers [*Ann. Chim. Phys.*, **18**, 5 (1909); *Compt. Rend.* **152**, 1380 (1911)] and others determined N_0 from Eq. 19-13. Boltzmann's constant was replaced by R/N_0. The gas constant R was known from work with gases. Their experiments are described by Glasstone [S. Glasstone, *Textbook of Physical Chemistry*, 2nd ed. (Macmillan and Co., London, 1953), p. 257]. They suspended granules of gamboge, a bright yellow gum resin, of density 1.194 in water at 25 C. Their granules had average radii of 0.368×10^{-4} cm. They found $\rho(h_2)/\rho(h_1) = 10^{0.481}$ for $h_2 - h_1$ of only 0.01 mm. What value of Avogadro's number did they determine?

19-7. Let a box of height h be filled with gas of molecular weight M in a constant gravitational field of acceleration g. What is the average height of molecules in the box?

19-8. The universal law of gravitation says that the gravitational force between two objects is $-m_1m_2G/r^2$; thus the gravitational potential is $-m_1m_2G/r$. Consider the equilibrium at temperature T between a planet of mass m_1 and molecules of mass m_2 in its atmosphere. Find the normalized gas density $\rho(r)$. Let d be the radius of the planet. Interpret the physical meaning of the result.

20 DENSE GASES AND LIQUIDS

INTERMOLECULAR POTENTIALS

The picture of the "ideal gas" is based on real gases in the limit of vanishing density. In fact, real gases have nearly ideal properties at all pressures up to critical if the temperature is at least two or three times the critical temperature. At lower temperatures or higher pressures, one must take into account the effects of the intermolecular interactions, as represented in Fig. 10-3. Since fully quantum mechanical treatment of these interactions is all but impossibly difficult, the intermolecular potential energy U of Eq. 17-3 is usually only approximated by an empirical function of the positions of the N particles that has the observed shape. This function is almost always taken to be a sum of independent contributions from each *pair* of particles in the system:

$$U_N(\mathbf{r}^N) = \sum_{\substack{i=1 \\ (i<j)}}^{N-1} \sum_{j=2}^{N} U_{ij}(r_{ij}) \equiv \sum_{i<j=1}^{N} U_{ij}(r_{ij}). \tag{20-1}$$

The second way of writing the sum is a brief equivalent to the first. The summation is over every pair of indices i and j. The restriction $i<j$ is imposed to assure the counting of each pair only once as U_{ij}, not again as U_{ji}. The pair potentials U_{ij} commonly look like those in Fig. 10-3, although particles in excited electronic states can have potentials with no attractive region at all. The approximation, Eq. 20-1, implies that the interaction potential between particles i and j does not change when particle k comes close to one or both of them, or even when particle k smacks one of them in a hard collision. Clearly, particle k's presence must distort the electron clouds of particles i and j, thus changing their interaction potential. Such effects are neglected by the use of Eq. 20-1. The direct interaction between particle k and the other two is of course still contained in Eq. 20-1 in the terms U_{ik} and U_{jk}. Probably at worst, use of Eq. 20-1 does no more than remove the sharp edge of accuracy from theoretical calculations of proper-

ties of dense gases or liquids. However, even using Eq. 20-1, a number of further approximations are usually needed. Furthermore, it is hard to reach agreement experimentally on the exact shapes even of the *pair* potentials U_{ij}. Thus it rarely happens that the pairwise additive intermolecular potential approximation is the most worrisome aspect of a particular problem.

In Eq. 20-1, U_{ij} is made a function of only the magnitude of the intermolecular distance $\mathbf{r}_{ij}$. This is acceptable for molecules that are more or less spherical but represents only an average for nonspherical molecules whose interaction depends also on the angles with which they are aligned.

Many different empirical representations of the pair potential are in common use. One of the simplest is the **hard sphere** model introduced in Chap. 18 ($U=\infty$ for $r \leqslant \sigma$; $U=0$ for $r>\sigma$). This approximation involves one adjustable parameter σ whose value is chosen to fit the substance at hand. This potential (plotted in Fig. 20-1*a*) gives an oversimplified account of the hard core repulsion between two molecules when they come too close together, but of course it ignores the intermolecular attraction at moderate distances of separation. This omission is acceptable when the temperature is so high that the depth of the potential well (ϵ in Fig. 20-1*b*) is much less than kT. For such high temperatures, the attractive wells represent simply minor perturbations in the collisions of what are for all practical purposes nearly hard spheres. The most commonly used two-parameter potential function is the so-called **Lennard-Jones 6–12 potential**, (Fig. 20-1*b*), whose mathematical form is

$$U(r)=4\epsilon\left[\left(\frac{\sigma}{r}\right)^{12}-\left(\frac{\sigma}{r}\right)^{6}\right]. \tag{20-2}$$

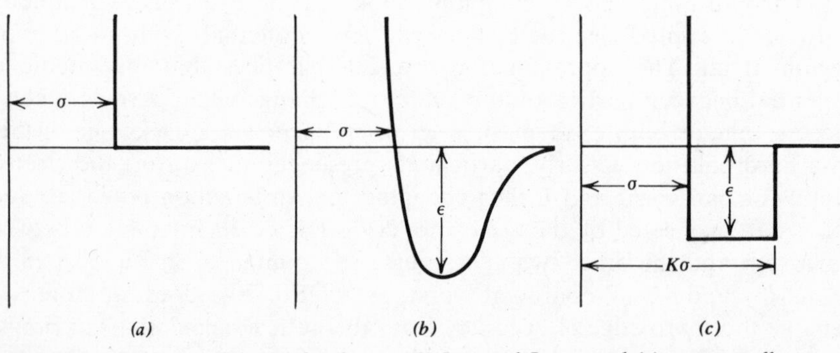

Fig. 20-1. Pair potentials: (*a*) hard core, (*b*) Lennard-Jones, and (*c*) square well.

The first term is a strongly repulsive core that drops off very quickly as r exceeds σ. The second term represents an attraction that falls off less quickly: it is canceled by the repulsion in the region of the core; but in the region about $r=\sigma$, where the first term is small, it forms the attractive well. By proper choice of ϵ and σ, the Lennard-Jones potential can be fit with considerable precision to a wide variety of substances. A three-parameter potential that is useful for its ease in calculations but clearly no improvement on the Lennard-Jones for accuracy is the **square well potential** of Fig. 20-1c:

$$\begin{aligned} U(r) &= \infty \qquad \text{for} \quad r \leqslant \sigma; \\ U(r) &= -\epsilon \qquad \text{for} \quad \sigma < r \leqslant K\sigma; \\ U(r) &= 0 \qquad \text{for} \quad r > K\sigma. \end{aligned} \tag{20-3}$$

For each pair of particles contributing U_{ij} to U_N, we arbitrarily say that *each particle* shares U_{ij} equally, thus attributing to each a potential energy of half U_{ij}. This means that the total potential energy associated with one particle (say the $(N+1)$st in a fluid containing a total of $N+1$ particles) is

$$u_{N+1}(\mathbf{r}^{N+1}) = \frac{1}{2} \sum_{i=1}^{N} U_{i,N+1}(\mathbf{r}^{N+1}). \tag{20-4}$$

It also means the addition of an $(N+1)$st particle to a fluid already containing N changes the energy by an amount

$$U_{N+1}(\mathbf{r}^{N+1}) - U_N(\mathbf{r}^N) = \sum_{i=1}^{N+1} U_{i,N+1}(\mathbf{r}^{N+1}) = 2u_{N+1}(\mathbf{r}^{N+1}). \tag{20-5}$$

The potential energy increase is twice the potential energy associated with the added particle; one of the u_{N+1}'s "belongs" to particle $N+1$ and the other "belongs" to the N original particles.

SEMICLASSICAL PARTITION FUNCTION

In classical mechanics the analogy to finding Q by summing $e^{-\beta E_{\text{state}}}$ over all states is the integration of $\exp[-\beta E(\mathbf{r}^N, \mathbf{p}^N)]$ over all phase space. This is shown in Eq. 17-2, and by using Eq. 17-3 the result will resemble the denominator of Eq. 17-4 (where we neglect internal states*):

$$\frac{1}{N!}\int d\mathbf{p}^N \exp\left(\frac{-\beta \sum_{i=1}^{N} p_i^2}{2m}\right)\int d\mathbf{r}^N e^{-\beta U_N(\mathbf{r}^N)}$$

$$=\frac{1}{N!}(2\pi mkT)^{3N/2}\int d\mathbf{r}^N e^{-\beta U_N(\mathbf{r}^N)}. \tag{20-6}$$

The momentum integrals were performed easily using Eq. B-1, but the important feature of Eq. 20-6 is the result that Q is proportional to Z, defined by

$$Z \equiv \int d\mathbf{r}^N e^{-\beta U_N(\mathbf{r}^N)}. \tag{20-7}$$

The quantity Z is often called the **configuration integral** (sometimes the quantity $Z/N!$ or Z/V^N or $Z/N!V^N$), since it is taken over the classical configuration space of the N particles.

The correct semiclassical partition function will contain Z as a factor, along with terms for translation, indistinguishability, and internal states. The easiest way to determine exactly what must multiply Z in the expression for Q is to compare Z and Q in the ideal gas limit. We found in Eq. 9-33 that the quantum mechanical value of Q for ideal gases is

$$Q_{\text{ideal gas}}=\frac{1}{N!}q^N. \tag{20-8}$$

The value of Z for ideal gases, where $U(\mathbf{r}^N)$ is zero, is simply V^N:

$$Z_{\text{ideal gas}}=V^N. \tag{20-9}$$

We thus can write Q in the form of something times Z for ideal gases:

$$Q_{\text{ideal gas}}=\frac{1}{N!}\left(\frac{q}{V}\right)^N V^N=\frac{1}{N!}\left(\frac{q}{V}\right)^N Z_{\text{ideal gas}}. \tag{20-10}$$

Since Z is the only part of Q that changes between real and ideal gases, the coefficient of Z in Eq. 20-10 must be the same regardless of whether the

*This implied independence of intermolecular potential and internal state is only an approximation, since rotational, vibrational, and electronic states do affect the potential. One usually hopes to be able to make ad hoc corrections for such difficulties whenever such changes appear to be important.

fluid is an ideal gas:

$$Q_{\text{semiclassical}} = \frac{1}{N!}\left(\frac{q}{V}\right)^N Z = \frac{1}{N!}\left(\frac{q}{V}\right)^N \int d\mathbf{r}^N e^{-\beta U_N(\mathbf{r}^N)}. \qquad (20\text{-}11)$$

The only modification semiclassical mechanics makes to the quantum mechanical partition function is to multiply it by Z/V^N. This permits classical incorporation of the effects of the intermolecular potentials; at the same time, the correction reduces to unity in the ideal gas limit. The value of q_{trans} was found in Eq. 10-18 to be $V(mkT/2\pi\hbar^2)^{3/2}$. Thus we write

$$Q_{\text{semiclassical}} = \frac{1}{N!}\left(\frac{mkT}{2\pi\hbar^2}\right)^{3N/2} q_{\text{int}}^N Z = \frac{1}{N!}\left(\frac{mkT}{2\pi\hbar^2}\right)^{3N/2} q_{\text{int}}^N \int d\mathbf{r}^N e^{-\beta U(\mathbf{r}^N)}.$$

(20-12)

It is interesting to compare Eq. 20-12 with the classical normalization integral, Eq. 20-6. The integral over N-particle phase space, Eq. 20-6, is $(2\pi\hbar)^{3N}$ or h^{3N} times as large as the sum over quantum states, Eq. 20-12. This suggests that a single N-particle quantum state is associated with a finite volume h^{3N} in phase space. This is indeed one form of the **Heisenberg uncertainty principle.** A small volume τ in phase space corresponds to approximately τ/h^{3N} quantum states.

THE CHEMICAL POTENTIAL

It is advantageous in studying dense fluids to work with the chemical potential, whose value is given by Eq. 8-20 as

$$\mu = -kT\ln\frac{Q_{N+1}}{Q_N} = -kT\ln\frac{[1/(N+1)!](q/V)^{N+1}Z_{N+1}}{1/N!(q/V)^N Z_N}, \qquad (20\text{-}13)$$

$$\mu = -kT\ln\frac{q}{NV}\frac{Z_{N+1}}{Z_N} = -kT\ln\frac{q\int d\mathbf{r}^{N+1} e^{-\beta U_{N+1}}}{NV\int d\mathbf{r}^N e^{-\beta U_N}}, \qquad (20\text{-}14)$$

where we replaced $N+1$ by N in the final result. The utility of working from Eq. 20-14 lies not only in its being correct, but primarily in its susceptibility to simple physical interpretation.

Thermodynamically, we note that Z_{N+1}/VZ_N is an expression for the *reciprocal of the activity, a,* of the fluid, where the standard state is chosen to be the hypothetical ideal gas at the same temperature and density as the real fluid. The proof is immediate: for the standard state ideal gas, Z_{N+1}/VZ_N in Eq. 20-14 is unity. Thus Eq. 20-14 is, in general,

$$\mu(\rho,T) \equiv \mu^0(\rho,T) - kT\ln a^{-1}(\rho,T) = \mu^0(\rho,T) - kT\ln\frac{Z_{N+1}}{VZ_N}, \quad (20\text{-}15)$$

which proves the assertion.

Now we consider two important intuitive ways to visualize Z_{N+1}/Z_N. We write this ratio as follows, taking advantage of Eq. 20-5:

$$\frac{Z_{N+1}}{Z_N} = \frac{\int d\mathbf{r}^N e^{-\beta U_N} \int d\mathbf{r}_{N+1} e^{-2\beta u_{N+1}}}{\int d\mathbf{r}^N e^{-\beta U_N}} = \frac{\int d\mathbf{r}_{N+1} \int d\mathbf{r}^N e^{-2\beta u_{N+1}} e^{-\beta U_N}}{\int d\mathbf{r}^N e^{-\beta U_N}}. \quad (20\text{-}16)$$

The first way of writing Eq. 20-16 shows (cf. Eq. 2-55) that Z_{N+1}/Z_N is the average value of the integral $\int d\mathbf{r}_{N+1} e^{-2\beta u_{N+1}}$ taken over the normalized probability density $d\mathbf{r}^N e^{-\beta U_N}/\int d\mathbf{r}^N e^{-\beta U_N}$. There are two equivalent ways to imagine taking this average: (1) first placing the $(N+1)$st particle and then averaging the N others throughout the volume, or (2) first placing the N other particles in a most likely configuration and then averaging the $(N+1)$st throughout the volume. Let us elaborate these approaches:

1. In finding the average in Eq. 20-16, we need to do the following: choose the position $\mathbf{r}_{N+1}$ of an $(N+1)$st particle and a small element of volume $d\mathbf{r}_{N+1}$ about $\mathbf{r}_{N+1}$. Then find the average of $e^{-2\beta u_{N+1}}$ in the presence of the N other particles that take up all different sets of positions characteristic of N particles in a volume V but do not "see" the $(N+1)$st particle. That means that they in no way adjust their positions because of the $(N+1)$st particle. So long as $\mathbf{r}_{N+1}$ is well away from a wall of the vessel, this average will be independent of choice of $\mathbf{r}_{N+1}$, and the integration over $d\mathbf{r}_{N+1}$ therefore will yield the factor V. Thus Eqs. 20-14 and 20-16 become

$$a^{-1} = \langle e^{-2\beta u}\rangle; \qquad \mu = -kT\ln\frac{q}{N}\langle e^{-2\beta u}\rangle, \quad (20\text{-}17)$$

where u is the potential energy of a particle fixed in a vessel containing N other particles, oblivious to the particle under consideration.

2. The second form of writing Eq. 20-16 suggests another, equivalent way of viewing this average. We recognize that the major contribution to the integrals over $d\mathbf{r}^N$ in numerator and denominator comes from a most probable distribution of distances of intermolecular separation. This distribution becomes overwhelmingly probable in the limit of an infinitely large system. If the distribution is taken to be the same in both the numerator and denominator, the volumes in N-particle configuration space and the $e^{-\beta U_N}$ terms for the most probable distributions cancel between numerator and denominator, leaving

$$\frac{Z_{N+1}}{Z_N} = \int d\mathbf{r}_{N+1} e^{-2\beta u_{N+1}}. \tag{20-18}$$

Here, the N particles are fixed in a most likely configuration and then the $(N+1)$st particle wanders throughout the entire system, each volume element being weighted by the exponential in the total intermolecular potential energy introduced by that particle.

These formal discussions will become clearer when the material is used to derive various practical results.

It is an exercise in thermodynamics to find the equation of state once an expression for a^{-1} is obtained as a function of density. We start with the Gibbs-Duhem equation for a single component at constant temperature:

$$d\mu = v\,dP = \frac{1}{\rho}dP \qquad \text{or} \qquad dP = \rho\,d\mu = \rho\left(\frac{\partial \mu}{\partial \rho}\right)_T d\rho. \qquad (T \text{ constant}) \tag{20-19}$$

From Eq. 20-15 we note

$$\left(\frac{\partial \mu}{\partial \rho}\right)_T = \left(\frac{\partial \mu}{\partial \rho}\right)_T^0 - kT\frac{\partial \ln a^{-1}}{\partial \rho}. \tag{20-20}$$

Using Eq. 20-20 in Eq. 20-19 and integrating yields

$$\frac{P}{\rho kT} = 1 - \frac{1}{\rho}\int_0^\rho \rho' \frac{\partial \ln a^{-1}}{\partial \rho'}d\rho' = 1 - \ln a^{-1} + \frac{1}{\rho}\int_0^\rho \ln a^{-1}(T,\rho')\,d\rho', \tag{20-21}$$

where the last of the two equivalent forms followed from the first on integration by parts. Thus knowing a^{-1} as a function of ρ from the dilute gas to the density of interest permits the calculation of the pressure at that density.

THE VIRIAL COEFFICIENTS

The **virial equation of state** for a gas is written to permit the expression of corrections to the ideal gas law as a power series in the density:

$$\frac{P}{\rho kT} = 1 + B(T)\rho + C(T)\rho^2 + D(T)\rho^3 + \cdots. \qquad (20\text{-}22)$$

The function $B(T)$ is called the **second virial coefficient,** $C(T)$ the **third virial coefficient,** and so on. (However, since the "first virial coefficient" turned out to be the gas constant R, or in the molecular case Boltzmann's constant k, people never call it by that name.) The virial coefficients are *functions of T only*—different functions, of course, for different pure substances or for mixtures of different composition.*

Finding how to calculate the first few virial coefficients from the shape of the intermolecular potential is surprisingly easy. Our approach will be to find the density expansion of $a^{-1} = Z_{N+1}/VZ_N$. It is shown in Prob. 20-1 that if a^{-1} is known in the form

$$a^{-1} = 1 + A_2\rho + A_3\rho^2 + A_4\rho^3 + \cdots, \qquad (20\text{-}23)$$

the virial coefficients are given immediately by

$$B = -\tfrac{1}{2}A_2; \qquad C = -\tfrac{2}{3}A_3 + \tfrac{1}{3}A_2^2; \qquad D = -\tfrac{3}{4}A_4 + \tfrac{3}{4}A_2A_3 - \tfrac{1}{4}A_2^3. \quad (20\text{-}24)$$

We will use the second interpretation of Z_{N+1}/Z_N, that of Eq. 20-18, to calculate the A coefficients. This means that we place the N particles in a most likely configuration for a dilute gas. Then, as the $(N+1)$st particle of Eq. 20-18 wanders throughout the system, the volume element $d\mathbf{r}_{N+1}$ is weighted by $e^{-2\beta u}$ in the neighborhood of other particles. Thus $d\mathbf{r}_{N+1}$ is weighted by unity in the part of the system in which it is far away from other particles. We approach this determination of Z_{N+1}/Z_N in a series of approximations, correct to higher and higher orders in the density. The first approximation is to ignore the presence of the N particles altogether. This is the ideal gas approximation, and the value obtained for Z_{N+1}/Z_N is V. Next we recognize that the N particles indeed are present in the system, but we approximate their positions as being separated from each other by distances of several times the range of the intermolecular forces. Consider then what happens when the $(N+1)$st particle comes close to particle i (which lies within $d\mathbf{r}_i$ with a probability $d\mathbf{r}_i/V$). In the ideal gas approximation we weighted $d\mathbf{r}_{N+1}$ in this part of the integral by 1. But

*See Andrews, pp. 120–121, 179.

clearly we should have weighted it by $e^{-\beta U_{i,N+1}}$. Therefore, we must introduce a correction term:

$$\frac{Z_{N+1}}{Z_N} \cong V + \sum_{i=1}^{N} \int d\mathbf{r}_{N+1} \int \frac{d\mathbf{r}_i}{V} (e^{-\beta U_{i,N+1}} - 1), \tag{20-25}$$

where we have counted a similar contribution from each of the N particles in the system. Since all N terms are identical, the value of the sum is N times the value of one term. The value of the quantity in the parentheses differs from zero only when the two particles are within the range of each other's forces. Thus so long as particle $N+1$ is a long way from a wall, we can change the integral over $d\mathbf{r}_i$ to one over the distance between the particles $d\mathbf{r}_{i,N+1}$ and can perform the integral over $d\mathbf{r}_{N+1}$ to yield the factor V. We finally are left with

$$a^{-1} = \frac{Z_{N+1}}{VZ_N} \cong 1 + \rho \int d\mathbf{r}_{i,N+1} (e^{-\beta U_{i,N+1}(r_{i,N+1})} - 1), \tag{20-26}$$

where we noted that the N terms in the sum were identical. In expressions like these it is customary for brevity to define the quantities e_{ij} and f_{ij}:

$$e_{ij} \equiv e^{-\beta U_{ij}}; \qquad f_{ij} \equiv e_{ij} - 1 \equiv e^{-\beta U_{ij}} - 1, \tag{20-27}$$

where both e_{ij} and f_{ij} are functions of intermolecular separation distance, as shown in Fig. 20-2. In terms of f_{ij}, the values of a^{-1}, $B(T)$, and $P/\rho kT$ are the following, a^{-1} and $P/\rho kT$ being correct to one power in ρ beyond the ideal gas:

$$\boxed{a^{-1} \cong 1 + \rho \int d\mathbf{r}_{ij} f_{ij}; \qquad B(T) = -\tfrac{1}{2} \int d\mathbf{r}_{ij} f_{ij}; \qquad \frac{P}{\rho kT} \cong 1 - \tfrac{1}{2}\rho \int d\mathbf{r}_{ij} f_{ij}.} \tag{20-28}$$

This equation of state, cut off (truncated) after the second virial term, is convenient to handle and is good even at low temperatures if the pressure is less than half the critical pressure. A very good approximation to the form of $B(T)$ is given by

$$B(T) \approx b - \frac{\alpha}{kT}, \tag{20-29}$$

where b and α are constants. The reason for this is easy to see from the form of B as $-\frac{1}{2}\int d\mathbf{r}_{ij} f_{ij}$. In the hard core region of the potential, f_{ij} is -1, as in Fig. 20-2. The integral over that region in the calculation of $B(T)$

thus yields

$$-\tfrac{1}{2}\int_{\text{hard core}} d\mathbf{r}_{ij} f_{ij} = \tfrac{1}{2}\int_{\text{hard core}} d\mathbf{r}_{ij} \cong b. \tag{20-30}$$

Thus b is just one-half the volume excluded to the center of one molecule by the presence of another molecule. This volume has almost no temperature dependence because the repulsive wall of the potential function rises so sharply with decreasing intermolecular separation. The hard core volume within which U is much greater than kT is almost completely independent of T (up to the point at which appreciable electronic excitation sets in). In the attractive region of the potential, however, an approximation must be made to perform the integration. If the results are limited to gases at moderately high temperature, the depth of the potential

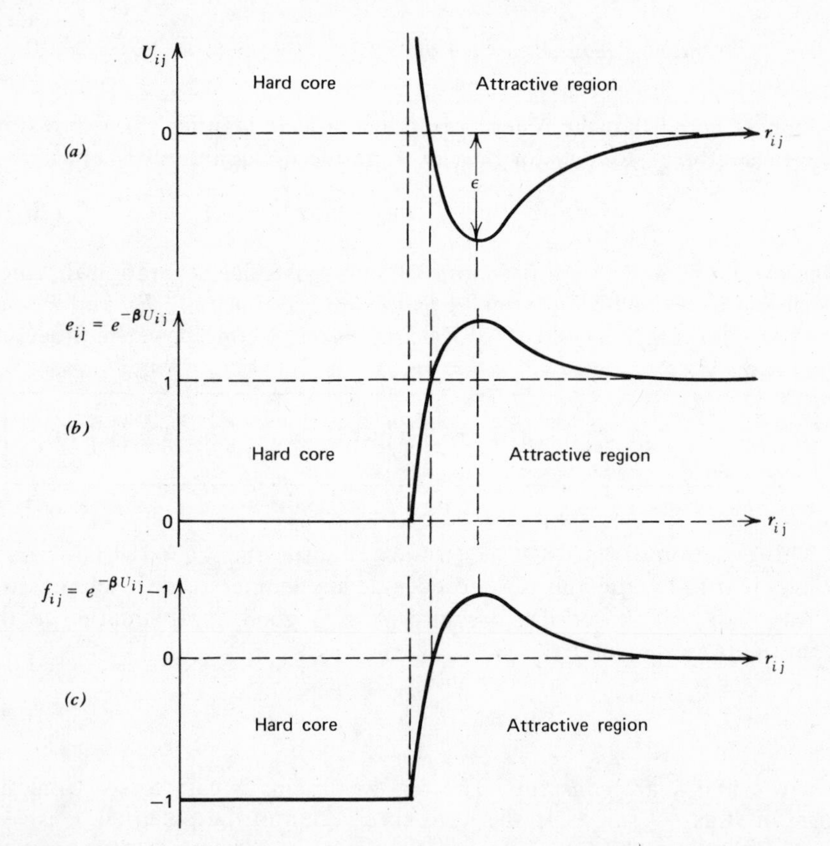

Fig. 20-2. Plots of useful functions of typical intermolecular potential U versus distance of separation r.

well (ϵ in either Fig. 20-1*b* or Fig. 20-2*a*) is less than kT. If it were greater, a number of particles, on colliding with others, would not have enough energy afterward to separate, thus quite likely leading to some degree of condensation of the gas into a liquid. Estimates of ϵ/k for several substances are presented in Table 20-1. For temperatures well above these

Table 20-1. Table of Ratios ϵ/k for Various Gases

Gas	Ne	Ar	Kr	N_2	CO_2	C_2H_6	C_3H_8	C_2H_4	CH_3Cl	NH_3	H_2O
ϵ/k K	20	70	98	54	119	244	347	222	470	692	1260

ϵ is the estimated depth of the intermolecular potential energy curve. Source: Hirschfelder, Curtiss, and Bird, p. 160.

values, the entire attractive region of the intermolecular potential is characterized by having $\beta U_{ij} \ll 1$. As a result of this, an expansion of the exponential in the integral for $B(T)$ in powers of βU can be stopped after the second term:

$$f_{ij} = e^{-\beta U_{ij}} - 1 = 1 - \beta U_{ij} + \tfrac{1}{2}\beta^2 U_{ij}^2 - \cdots - 1 \cong -\beta U_{ij}. \qquad (20\text{-}31)$$

The constant β is not involved in the integral over the attractive region, and this integral leaves a negative constant, $-\alpha$, times β:

$$-\frac{1}{2}\int_{\text{attractive region}} d\mathbf{r}_{ij} f_{ij} \cong \frac{1}{2kT}\int_{\text{attractive region}} d\mathbf{r}_{ij}\, U_{ij}(r_{ij}) = -\frac{\alpha,}{kT}. \qquad (20\text{-}32)$$

Thus the α of Eq. 20-32 is just half the integral over the volume of the attractive region of the magnitude of the intermolecular potential in that region. These results suggest one way of learning about the shapes of the intermolecular potentials U_{ij} by simply measuring $B(T)$ for the gas as a function of temperature. This is considered further in Prob. 20-3.

Expressing $C(T)$ and even $D(T)$ in terms of similar integrals over the potential is straightforward with this method.* Expressions for higher virial coefficients, of course, become more complicated.

---

*Let us refine Eq. 20-25 to account for the pairs among the N particles which lie close together. The joint probability that particle i lies in $d\mathbf{r}_i$ and particle j lies in $d\mathbf{r}_j$ is $d\mathbf{r}_i d\mathbf{r}_j e_{ij}/V^2$, to lowest order in density. If they do lie there, the $d\mathbf{r}_{N+1}$ that contributed to the V term in Eq. 20-25 must be changed to $e_{i,N+1}e_{j,N+1}d\mathbf{r}_{N+1}$. In addition, the contributions of particles i and j to the sum in Eq. 20-25 must be subtracted, since their contributions are

ONE–DIMENSIONAL GAS OF HARD LINES

For gases having less than half the critical pressure, the virial expansion is easy to handle mathematically and is directly related to the intermolecular potentials through Eq. 20-28 (for $T > T_{\text{crit}}$ the pressure may be as high as P_{crit}). However, at higher pressures or at liquid densities the virial expansion is no good and one must find something else. No very satisfactory statistical mechanical theory has been developed for dense fluids, but we will spend the rest of this chapter exploring some interesting approaches.

First, because of its great simplicity and what it will suggest for three-dimensional fluids, let us visualize a hypothetical *one-dimensional* fluid in which N molecules are constrained to move along a line of length L, rather than in a box of volume V. Virtually the only change introduced by this new fluid is the replacements in Eqs. 20-11 and 20-12 of V by L and of the exponents $\frac{3}{2}N$ by $\frac{1}{2}N$. The configuration integral in one dimension is

$$Z_N = \int_0^L dl^N e^{-\beta U_N(l^N)}, \tag{20-35}$$

and the ratio Z_{N+1}/Z_N, thus the expression for a^{-1}, is analogous to the result found earlier for three dimensions.

In particular, let us calculate a^{-1} for a gas of *one-dimensional hard lines* or rods of length σ, particles whose potential is analogous to the hard spheres of Fig. 20-1a except r_{ij} is replaced by l_{ij}. We fix N particles on the length L in a most likely configuration (see Fig. 20-3). Since there is no attractive potential well to correlate the distances between particles, they must not overlap but will otherwise have random separation distances. The value of Z_{N+1}/Z_N from Eq. 20-18 is simply the total length along which

---

given already. The result is the correction term

$$V^{-2} \sum_{i<j=1}^{N} \int d\mathbf{r}_{N+1} d\mathbf{r}_i d\mathbf{r}_j e_{ij}[(e_{i,N+1}e_{j,N+1} - 1) - (e_{i,N+1} - 1) - (e_{j,N+1} - 1)]$$

$$\equiv V^{-2} \sum_{i<j=1}^{N} \int d\mathbf{r}_{N+1} d\mathbf{r}_i d\mathbf{r}_j e_{ij} f_{i,N+1} f_{j,N+1} \tag{20-33}$$

to Z_{N+1}/Z_N or

$$A_3\rho^2 = \tfrac{1}{2}\rho^2 \int d\mathbf{r}_{in} d\mathbf{r}_{jn} e_{ij} f_{in} f_{jn} \tag{20-34}$$

to a^{-1}. In terms of this, Eq. 20-24 permits calculating the third virial coefficient. Statistical mechanical expressions for additional virial coefficients are obtained by extending this procedure.

the $(N+1)$st particle could move without overlapping one of the N fixed particles. In Fig. 20-3 this is simply the sum of the lengths of the arrows drawn below the line. This is the same as L times the probability that a point picked at random in the box of length L lies sufficiently far from particles on both sides to accommodate another particle. This probability is precisely what is meant by $\langle e^{-2\beta u}\rangle$ in Eq. 20-17, since u is either infinity or zero for this potential. Thus the two approaches of Eqs. 20-17 and 20-18 give identical results.

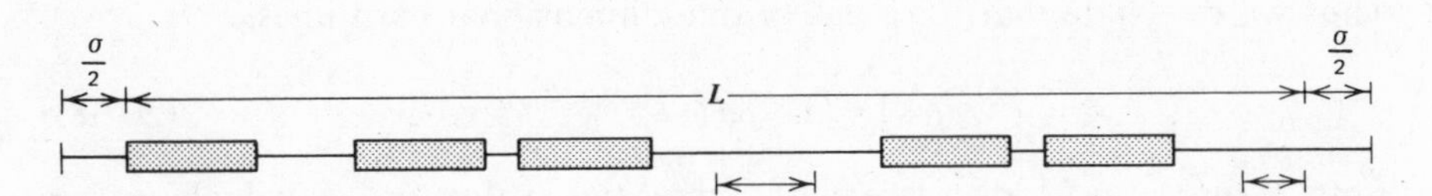

Fig. 20-3. One-dimensional box of length approximately L, containing N hard line molecules of length σ.

We shall compute this probability as the product of two terms: the first is the probability that a point x picked at random between 0 and L lies in a hole between two adjacent hard line molecules (i.e., that the center of no other particle lies between $x-\frac{1}{2}\sigma$ and $x+\frac{1}{2}\sigma$). From Fig. 20-3 it is clear that this is simply the fraction $(1-\rho\sigma)$ of the box that is unoccupied by hard lines. The second term is the probability that the point is the center of a hole of length σ or more (i.e., the probability that, conditional on the first, no other particle lies between $x-\sigma$ and $x+\sigma$). The second term can be rephrased as follows: consider the distribution of distances from one edge of a particle to the next (i.e., the way in which the total empty space $L-N\sigma$ is divided among the N interparticle separations). This distribution will be the same as the distribution of separation distances between N points placed randomly on a line of length $L-N\sigma$ (see Fig. 20-4). If N points are placed randomly on a line of length $L-N\sigma$, what is the probability that no such point lies within a particular region of length σ, where $\sigma \ll L$? This is easily calculated, since each successive point has a chance $1-\sigma/(L-N\sigma)$ of being outside the region of length σ. Thus the

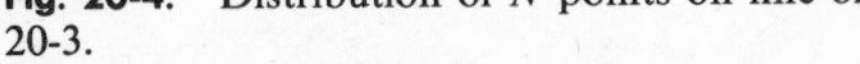

Fig. 20-4. Distribution of N points on line of length $L-N\,\sigma$; completely equivalent to Fig. 20-3.

probability Π that all N of them lie outside it is

$$\Pi=\left(1-\frac{\sigma}{L-N\sigma}\right)^N. \tag{20-36}$$

Since $\sigma \ll L-N\sigma$, the limiting form of Eq. 20-36 as N goes to infinity is the well-known formula (see Prob. 20-4):

$$\lim_{N\to\infty} \Pi=e^{-N\sigma/(L-N\sigma)}=e^{-\rho\sigma/(1-\rho\sigma)}. \tag{20-37}$$

Thus we conclude that for a gas of one-dimensional hard lines,

$$a^{-1}=(1-\rho\sigma)e^{-\rho\sigma/(1-\rho\sigma)}. \tag{20-38}$$

Both terms in a^{-1} decrease with increasing value of $\rho\sigma$, which is the fraction of the box length occupied by the N molecules. This shows how the activity itself increases with increasing value of $\rho\sigma$, thus with decreasing unoccupied or free volume.

In Prob. 20-5 we use Eq. 20-21 to demonstrate that the equation of state for the hard line gas that follows from Eq. 20-38 is*

$$\boxed{P=\frac{\rho kT}{1-\rho\sigma}.} \tag{20-39}$$

The sole difference between this and the ideal gas is the denominator of $1-\rho\sigma$ rather than 1. This is identical to the hard core term in van der Waals' equation of state for dense gases.

ONE–DIMENSIONAL VAN DER WAALS EQUATION

Suppose now we add to our one-dimensional hard lines an attractive potential such that (*a*) it does not perturb the positions of the particles from what they would have as hard lines of the same density, and (*b*) the total attractive potential energy experienced on average by a particle is simply proportional to the density of particles ρ. Such a potential well must be very shallow with respect kT to fulfill condition *a*, and it must be very long range to fulfill condition *b*. An example would be the square well

*This result was originally obtained by L. Tonks, *Phys. Rev.*, **50**, 955 (1936), by a more complicated procedure.

potential of Eq. 20-3 and Fig. 20-1c, where $\epsilon/kT \ll 1$ and $K\sigma \gg L/N$. Since the particles' positions are unperturbed, the hard core effect on a^{-1} is unchanged from Eq. 20-38. The only difference is that the $(N+1)$st particle always experiences an average attractive potential energy of $-\rho(K-1)\sigma\epsilon$ from being within the attractive wells of all the $2\rho(K-1)\sigma$ particles within range. Thus the sole effect of the attractive wells is to change a^{-1} from the value given by Eq. 20-38 to

$$a^{-1} = (1-\rho\sigma)e^{-\rho\sigma/(1-\rho\sigma)}e^{\rho(K-1)\sigma\epsilon/kT}. \tag{20-40}$$

Here we can see quantitatively how the hard cores and the attractive wells of the molecules work at cross purposes in affecting the value of the activity (which thermodynamics shows to be a measure of the driving force for diffusive flow of matter). Hard cores tend to force particles out of a system, thus increasing the activity. Attractive wells tend to hold particles into the bulk material, decreasing the activity.

In Prob. 20-6 it is shown that the equation of state that follows from Eq. 20-40 by the use of Eq. 20-21 is

$$\boxed{P = \frac{\rho kT}{1-\rho\sigma} - \frac{1}{2}\rho^2(K-1)\sigma\epsilon,} \tag{20-41}$$

which is the **van der Waals equation** in one dimension, where σ is the customary b parameter, and $\frac{1}{2}(K-1)\sigma\epsilon$ the usual a parameter.* We thus have found the two criteria a and b above, under which a one-dimensional fluid with intermolecular potentials of hard cores and attractive wells will obey the van der Waals equation.

Although it does not fall directly out of the statistical mechanics, it is clear that Eqs. 20-40 and 20-41 describe a fluid that exists both as a gas and as a liquid at temperatures below a critical temperature of $8(K-1)(\epsilon/27k)$ (see Prob. 20-7). If the entire fluid region—vapor, liquid, and gas—is described by those equations, the phase coexistence region is determined by the twin requirements that $P_{\text{liq}} = P_{\text{vap}}$ and $\mu_{\text{liq}} = \mu_{\text{vap}}$ for liquid and vapor that coexist. This is explore further in Prob. 20-8.

*This result was originally obtained through a complicated analysis by G. E. Uhlenbeck, M. Kac, and P. C. Hemmer, *J. Math. Phys.*, **4**, 216 and 229 (1963).

THREE–DIMENSIONAL VAN DER WAALS EQUATION

The van der Waals equation therefore is justified theoretically for one-dimensional fluids whose intermolecular potential is a hard core plus a very shallow, long-range attractive well. However, regardless of the number of dimensions a system has, the preceding section clearly indicates that if the attractive wells do not alter the relative positions of the particles from what the hard sphere fluid would have at the same density, then

$$a^{-1} = a^{-1}_{\text{hard cores}} a^{-1}_{\text{attr wells}}, \qquad \text{where } a^{-1}_{\text{attr wells}} = \langle e^{-2\beta u_{\text{attr}}} \rangle. \tag{20-42}$$

Here, the average is that felt by a test particle moving throughout the free volume between the hard cores of N typically placed molecules. Use of Eq. 20-42 in Eq. 20-21 shows the result of this independence of hard cores and attractive wells on the equation of state:

$$P = P_{\text{hard cores}} + P_{\text{attr wells}} = P_{\text{hard cores}} - \frac{1}{\rho}\int_0^{\rho} \rho' \frac{\partial}{\partial \rho'} \langle e^{-2\beta u_{\text{attr}}} \rangle \, d\rho'. \tag{20-43}$$

Since the attractive wells obviously affect the ways in which molecules clump together in the fluid, this form is not rigorous. It is, however, a tolerable approximation, perhaps better for liquids than for dense gases, since liquids are so dense that the molecules go almost every place there is room, a feature determined by the hard cores alone. The approximation of Eq. 20-43 is no good at all in the vicinity of the critical point.

Again, regardless of the number of dimensions, the exact form of van der Waals' attractive term arises from the additional assumption that the average attractive potential energy experienced by a molecule wandering about between the others in the fluid is proportional to ρ:

$$a^{-1}_{\text{attr wells}} = e^{\rho \Lambda \epsilon / kT}, \tag{20-44}$$

where Λ is the effective volume of the attractive well and ϵ the average well depth within that volume. For the square well potential, ϵ is the depth of the well and

$$\Lambda = \tfrac{4}{3}\pi (K^3 - 1)\sigma^3. \tag{20-45}$$

The resulting equation of state is

$$P = P_{\text{hard cores}} - \tfrac{1}{2}\rho^2 \Lambda \epsilon, \tag{20-46}$$

in analogy to Eq. 20-41. The assumption involved here is valid at low density, but only qualitatively reasonable at higher densities for realistically short-range potentials. The higher the density, the more of the time the $(N+1)$st particle will be in the attractive well of one or more neighbors. One objection to the van der Waals attractive term is its high-density limit. As is discussed in the next section, compressed liquid densities are of the order of $\rho \approx 1.1\sigma^{-3}$, for which the value of $\rho\Lambda$ is $4.6(K^3-1)$. This is the average number of potential wells in which a typical molecule will lie at this high density (i.e., the coordination number of the molecules in the liquid). Since values of K vary from 1.2 for steam, to 1.85 for the rare gases, to 2.3 for $CHCl_2F$, the van der Waals coordination numbers for these liquids would range from 3.3 to 25 to 51, respectively. In defense of such values one can say that the longer the range of the attractive well, the more molecules beyond just its nearest neighbors will be felt by a given particle. A molecule with 8 to 12 nearest neighbors may well have dozens of molecules beyond those that are still within range of its attractive forces. Viewed in that light, these coordination numbers are perhaps less undesirable.

Unfortunately, in three dimensions, the problem of finding the volume available to the $(N+1)$st hard sphere as it moves about in the system is far more complicated than the counterpart problem in one dimension that led to Eq. 20-38. In fact, there is no justification for the form of Eq. 20-38 in three dimensions except that it worked in one dimension. Thus it is not surprising that van der Waals' equation is found to fit experimental data only in a very restricted range of volumes and pressures.

A NEW TREATMENT OF THREE-DIMENSIONAL HARD SPHERE FLUID

Let us attempt to understand three-dimensional fluids by approximating first the hard spheres, and adding on van der Waals attractive wells. There are a number of statistical mechanical approaches to hard spheres,* but most of them are extremely complicated or extremely inaccurate (or, usually, both). The approach offered here is at least simple, and its precision seems to be about as good as the best of the complicated theories.

*For surveys see G. S. Rushbrooke, in *Physics of Simple Liquids*, H. N. V. Temperley, J. S. Rowlinson, and G. S. Rushbrooke, eds., (Wiley-Interscience, New York, 1968), Chap. 2; or see S. A. Rice and P. Gray, *The Statistical Mechanics of Simple Liquids* (Wiley-Interscience, New York, 1965), Chap. 2. The most precise of these theories is probably that of H. Reiss, H. L. Frisch, and J. L. Lebowitz, *J. Chem. Phys.*, **31**, 369 (1959); and also of J. K. Percus and G. J. Yevick, *Phys. Rev.* **110**, 1 (1958); E. Thiele, *J. Chem. Phys.*, **39**, 474 (1963); M. S. Wertheim, *Phys. Rev. Lett.*, **10**, 321 (1963), *J. Math. Phys.*, **5**, 643 (1964). An improved version of the material presented in this section is given in F. C. Andrews, *J. Chem. Phys.*, in press (1974).

We know that a^{-1} is the probability that a point chosen at random in the system containing N particles in a most likely configuration has no neighboring particle with its center within a distance σ. Again, as we did in one dimension, we compute this as the product of two terms. The first term is the probability that a point $\mathbf{r}$ chosen at random within V does not lie inside the core of one of the N molecules. This we know to be simply the fraction

$$1-\frac{4}{3}\pi\left(\frac{\sigma}{2}\right)^3\rho=1-\frac{\pi}{6}\sigma^3\rho\equiv 1-\frac{1}{8}b\rho \tag{20-47}$$

of the space that is not occupied by hard cores. For simplicity we have defined the quantity b to be the volume $\frac{4}{3}\pi\sigma^3$ excluded by an isolated molecule of diameter σ, thus $\frac{1}{8}b$ is the actual volume of each sphere. The second term we compute is the probability that about $\mathbf{r}$, already known to have no particle within a distance of $\frac{1}{2}\sigma$, there is also no particle within σ (i.e., within the additional volume $b-\frac{1}{8}b=\frac{7}{8}b$ about $\mathbf{r}$). We take the probability Π that all N molecules lie outside this new volume of $\frac{7}{8}b$ to be the Nth power of the probability that a particular one lies outside it, in analogy to Eq. 20-36. In the one-dimensional case, this individual probability was $\sigma/(L-N\sigma)$, the ratio of the volume of interest to the total free volume. This total free volume $L-N\sigma$ is what would be available if all the molecules crowded into one side of the box. The analogy in three dimensions is the ratio of $\frac{7}{8}b$ to the free volume $V-N\omega$, where ω is the average volume taken up by each molecule when they are crowded as tightly as possible into one side of the box. We leave the value of ω unspecified for the moment and simply write the three dimensional analog to Eqs. 20-36 and 20-37:

$$\lim_{N\to\infty}\Pi=\left(1-\frac{7b}{8(V-N\omega)}\right)^N=e^{-7Nb/8(V-N\omega)}=e^{-7\rho b/8(1-\rho\omega)}. \tag{20-48}$$

Thus for a three-dimensional hard sphere fluid we obtain

$$\boxed{a^{-1}=(1-\tfrac{1}{8}\rho b)e^{-7\rho b/8(1-\rho\omega)}.} \tag{20-49}$$

In Prob. 20-10 we use Eq. 20-21 to show that the equation of state that follows from Eq. 20-49 is

$$\boxed{\frac{P}{\rho kT}=\frac{7b}{8\omega(1-\rho\omega)}+\frac{7b}{8\omega^2\rho}\ln(1-\rho\omega)-\frac{8}{\rho b}\ln(1-\tfrac{1}{8}\rho b).} \tag{20-50}$$

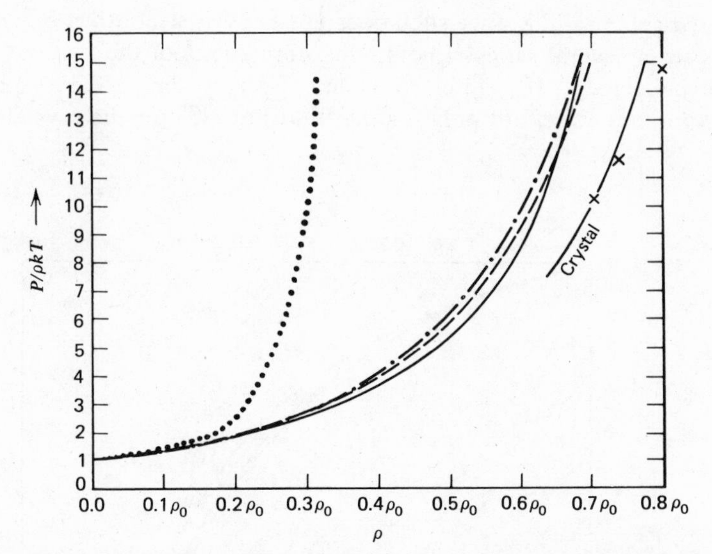

Fig. 20-5. Plots of $P/\rho kT$ versus ρ, where ρ_0 is density of close-packed crystal, $1.414\sigma^{-3}$. Solid curve, Eq. 20-50 with $\omega = 0.8225\ \sigma^3$ for fluid and $\omega = 0.7071\ \sigma^3$ for crystal; dashed curve, molecular dynamics result; dashed and dotted curve, Reiss, Frisch, Lebowitz–Percus Yevick equation; dotted curve, van der Waals equation with b adjusted to give correct second virial coefficient; crosses are molecular dynamics crystal data.

We now must establish the value of ω, the average volume occupied by a sphere when all spheres are crushed together into one end of the box and have no molecule-sized holes between them. There are two possible ways they could crowd together: in a random, disorganized manner, or in a cooperative, ordered, close-packed structure resembling that of a crystal. In the latter case, the calculation of ω is straightforward; its value is found in Prob. 20-9 to be $0.7071\sigma^3$. In the former case, the value of ω has been determined from a large number of experiments with actual spheres* to be approximately $0.8225\sigma^3$, a value whose calculation by purely theoretical means would be of considerable interest. Figures 20-5 and 20-6 plot $P/\rho kT$ and $\ln a^{-1}$ against density using Eq. 20-50, for the two possible choices of ω. Also shown are values similar to those given by the van der Waals equation and by the Reiss, Frisch, Lebowitz–Percus Yevick equation. The latter is considered to be the most precise of the rather complicated statistical mechanical treatments of hard spheres. The "experimental values" with which these results are compared are of course not those for a

*J. D. Bernal, in *Liquids; Structure, Properties, Solid Interactions*, T. J. Hughel, ed., (Elsevier Publishing Co., Amsterdam, 1965), p. 25; G. D. Scott and D. M. Kilgour, *Brit. J. Appl. Phys.* (*J. Phys. D*), ser. 2, **2**, 863 (1969); J. L. Finney, *Proc. Roy. Soc. London A*; **319**, 479 (1970).

real hard sphere fluid, since such does not exist. Instead, they represent the results from several series of computer simulations of the motions of hard sphere molecules.* It is interesting that the very simple theory presented here is quite precise, not only for the fluid but also for the crystal.

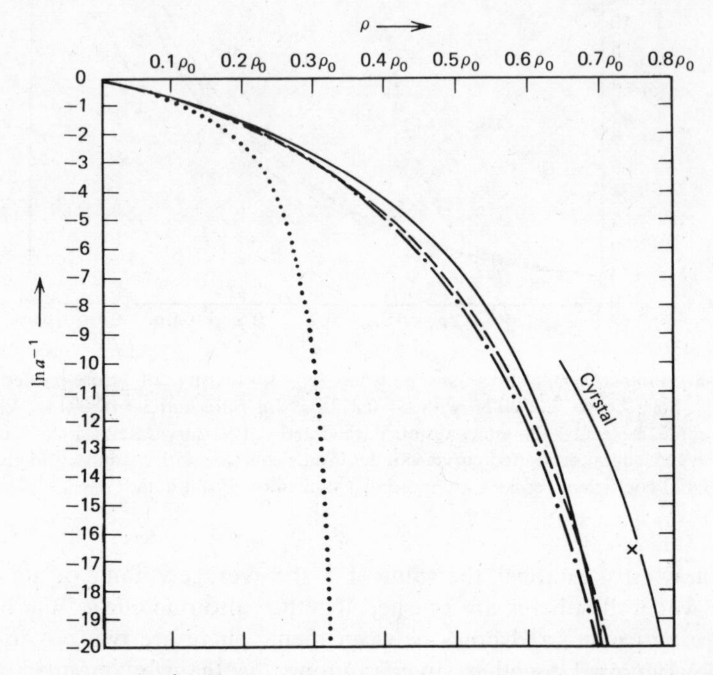

Fig. 20-6. Plots of ln a^{-1} versus ρ where ρ_0 is density of close-packed crystal, 1.414 σ^{-3}. Solid curve, Eq. 20-49 with $\omega = 0.8225\sigma^3$ for fluid and $\omega = 0.7071\sigma^3$ for crystal; dashed curve molecular dynamics result; dotted and dashed curve, Reiss, Frisch, Lebowitz–Percus Yevick equation; dotted curve van der Waals equation with b adjusted to give correct second virial coefficient; × is point calculated for crystal from molecular dynamics data.

A great deal of evidence suggests that it is indeed the hard cores of real molecules that goven the structures of real liquids, thus their properties. For example, consider a possible crystal-liquid phase transition in Fig. 20-6. The criterion of the equality of chemical potentials leads to an

*For the fluid, these are from B. J. Alder and T. E. Wainwright, *J. Chem. Phys.*, **33**, 1439 (1960), as represented analytically by N. F. Carnahan and K. E. Starling, *J. Chem. Phys.*, **51**, 635 (1969). For the crystal, these are from B. J. Alder, W. G. Hoover, and D. A. Young, *J. Chem. Phys.*, **49**, 3688 (1968).

equality of $\ln a^{-1}$ for the two phases (to within about 0.1). Thus typical equilibrium densities are $0.75\rho_0$ and $0.67\rho_0$, which represents a 12% increase in volume on melting. A host of different simple substances show precisely that—a volume increase of about 14 to 15% on melting. Of course there is a 15% volume increase in the values of ω, $0.7071\sigma^3$ and $0.8225\sigma^3$, at which the two phases are as closely packed as is possible.

APPROXIMATE TREATMENT OF THREE-DIMENSIONAL FLUID

We now consider a fluid of real molecules, hard cores plus attractive wells, and assume that the hard cores are the predominant influence on the molecules' positions. This is reasonable at both low and high density although at intermediate densities the clumping caused by the attractive forces would demand a more sophisticated treatment. For the square well potential, the results of Eqs. 20-42 to 20-46 are then valid, and our approximate treatment yields the following results:

$$\ln a^{-1} = \ln\left(1 - \frac{\pi\sigma^3\rho}{6}\right) - \frac{7\pi\sigma^3\rho}{6(1-\omega\rho)} + \frac{4\pi(K^3-1)\epsilon\sigma^3\rho}{3kT}; \qquad (20\text{-}51)$$

$$\frac{P}{\rho kT} = \frac{7\pi\sigma^3}{6\omega(1-\omega\rho)} + \frac{7\pi\sigma^3}{6\omega^2\rho}\ln(1-\omega\rho) - \frac{6}{\pi\sigma^3\rho}\ln\left(1-\frac{\pi\sigma^3\rho}{6}\right) - \frac{2\pi(K^3-1)\epsilon\sigma^3\rho}{3kT}. \qquad (20\text{-}52)$$

We have used Eq. 20-44 with 20-49 and Eq. 20-46 with 20-50. For fluids, the value of ω in Eqs. 20-51 and 52 is $0.8225\sigma^3$, for crystals $\omega = 0.7071\sigma^3$.

The most studied of all simple fluids is argon, for which the best values of the square well potential parameters as determined from second virial coefficient data are $\sigma = 3.067$ Å, $K = 1.70$, $\epsilon/k = 93.3$ K.* For these values,

*A. E. Sherwood and J. M. Prausnitz, *J. Chem. Phys.*, **41**, 429 (1964).

Eqs. 20-51 and 20-52 become

$$\ln a^{-1} = \ln(1-9.09d) - \frac{63.7d}{(1-14.28d)} + \frac{26{,}560d}{T}, \qquad \text{(Ar, fluid)} \qquad (20\text{-}53)$$

$$\ln a^{-1} = \ln(1-9.09d) - \frac{63.7d}{(1-12.28d)} + \frac{26{,}560d}{T}, \qquad \text{(Ar, crystal)} \quad (20\text{-}54)$$

$$\frac{P}{\rho kT} = \frac{4.46}{(1-14.28d)} + \frac{0.312}{d}\ln(1-14.28d)$$

$$- \frac{0.1100}{d}\ln(1-9.09d) - \frac{13{,}280d}{T}, \qquad \text{(Ar, fluid)} \quad (20\text{-}55)$$

$$\frac{P}{\rho kT} = \frac{5.18}{(1-12.28d)} + \frac{0.422}{d}\ln(1-12.28d)$$

$$- \frac{0.1100}{d}\ln(1-9.09d) - \frac{13{,}280d}{T}, \qquad \text{(Ar, crystal)} \quad (20\text{-}56)$$

where d is the density in moles per cubic centimeter.

In Fig. 20-7 we plot PVN_0/N, which is the product of pressure and molar volume or just RT times Eq. 20-55, against density for three constant temperatures, and compare with similar experimental plots. Comparisons such as these are the crucial test of any theory of liquids. Considering how many assumptions are built into our theory and how very approximate the square well potential is, the theory fits experiment surprisingly well. Another experimental check on the theory is to predict the critical point through the usual procedure of setting $(\partial P/\partial\rho)_T = 0$ and $(\partial^2 P/\partial\rho^2)_T = 0$ (Prob. 20-15). For argon, our simple theory predicts $d_{\text{crit}} = 0.0158$ mole/cm^3 and $T_{\text{crit}} = 147.5$ K. The experimental values are $d_{\text{crit}} = 0.0134$ mole/cm^3 and $T_{\text{crit}} = 150.9$ K, which deviate by 15% and 2%, respectively, from the calculations.

Our treatment of liquids will have to stop here, even though the work is incomplete. Liquids are one of the three major states of matter and are indispensable to life, but they are not well understood. They involve too many mutually interacting molecules to be as simple as gases, and they lack the molecular order that permits the treatment of crystals. The particular theoretical approach of the last part of this chapter may never prove to be as satisfying as one would like. At least it has the virtues of

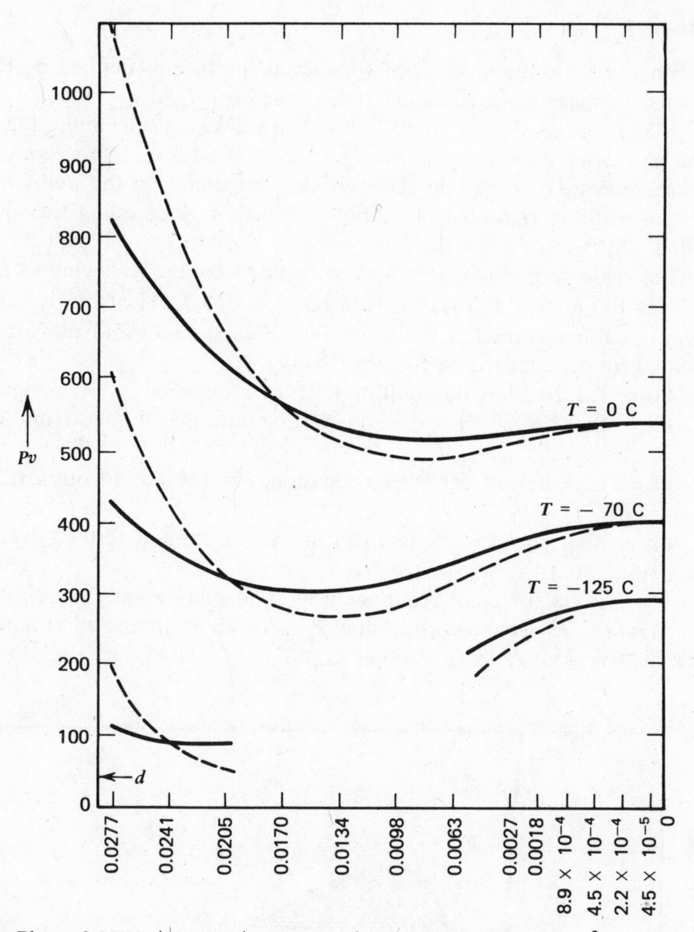

Fig. 20-7. Plots of $PVN_0/N = PV/n = Pv$, cal/mole, versus d, mole/cm^3 for argon at $T=0$ C, -70 C, and -125 C. Solid curve generated by Eq. 20-55; dashed curves, experimental results of A. Michels, J. M. Levelt, and W. de Graaff, *Physica*, **24**, 659 (1958); d-axis is expanded at low density.

being relatively simple mathematically, relatively intuitive physically, and fairly accurate. Other theories combining those virtues in different combinations exist, and still others will be developed. Perhaps some day, as people continue to add to our knowledge, this fascinating problem will be solved and we will feel that we truly understand the liquid state. Perhaps, but maybe not.

PROBLEMS

20-1. Prove the results of Eq. 20-24 for virial coefficients defined by Eq. 20-22 in terms of the expansion coefficients of Eq. 20-23 for a^{-1}.

20-2. Many statistical mechanics books have a drawing resembling Fig. 20-2, in which the plot of e_{ij} goes along at zero and then slowly, smoothly, with a round corner, turns upward toward one. This effect is attributed to the well-known fact that U_{ij} does not rise rigorously to infinity. Which way of doing the drawing is correct, and why?

20-3. For neon, experimental values of $N_0B(T)$ are the following as functions of T: $-7\frac{1}{2}$ at 100 K, 0 at 123 K, 6 at 175 K, $9\frac{1}{2}$ at 235 K, 11 at 273 K, $13\frac{1}{2}$ at 373 K, all data in cubic centimeters per mole. Plot B versus $1/T$ and suggest a square well potential for Ne that would fit these data.

20-4. Prove Eq. 20-37 as the limiting form of Eq. 20-36.

20-5. Prove that Eq. 20-39 is the equation of state that follows from Eq. 20-38 for a^{-1}.

20-6. Prove that the van der Waals equation, 20-41, follows from Eq. 20-40 for a^{-1}.

20-7. Prove that the critical temperature for a fluid described by van der Waals' equation, 20-41, is $8(K-1)\epsilon/27k$.

20-8. The criteria for equilibrium between two phases are $P(\rho_1,T)=P(\rho_2,T)$ and $\mu(\rho_1,T)=\mu(\rho_2,T)$. Suppose you know $P=P(\rho,T)$ as an analytic function over the entire fluid region. A single isotherm at $T<T_{\text{crit}}$ looks like Fig. 20-8. Prove

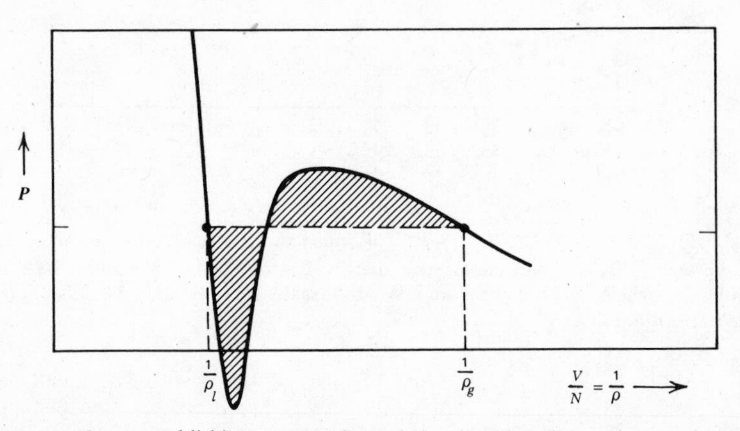

Fig. 20-8. Isotherm establishing nature of coexisting liquid and gas phases through use of Maxwell's equal area rule.

Maxwell's equal area rule, that the value of P and values of ρ which represent coexisting phases are those which make the two cross-hatched areas equal.

20-9. Prove that the value of ω for close-packed spheres is indeed $0.7071\sigma^3$.

20-10. Prove that Eq. 20-50 is indeed the equation of state that follows from Eq. 20-49 for a^{-1}.

20-11. Scott and Kilgour give the value 0.6366 ± 0.0005 as the random close-packed density of hard spheres. The density is the fraction of the space that is actually occupied by the spheres. Show that the average volume per sphere under such conditions is indeed approximately $0.8825\sigma^3$. What is the crystalline close-packed density?

20-12. The temperature at which $B(T)=0$ is often called the **Boyle temperature**. Why?

20-13. Argon has to be raised to a much higher temperature than neon before its $B(T)$ becomes almost independent of T, and helium's $B(T)$ is almost T-independent at room temperature. Why?

20-14. Verify Eqs. 20-53 to 20-56.

20-15. Verify the values of the critical density and temperature obtained for argon from Eq. 20-55 or 20-53 as quoted in the text.

APPENDIX A: REFERENCES

d'Abro, A., *The Rise of the New Physics*, D. Van Nostrand Co., 1939, reprinted in paperback in two volumes by Dover Publications, New York, 1951. A comprehensive work on the development of modern physics, especially quantum mechanics, thermodynamics, and statistical mechanics. Written for the intelligent student; long on physical intuition and mathematical conclusions; a book too little read these days.

Andrews, F. C., *Thermodynamics—Principles and Applications*, Wiley-Interscience, New York, 1971. A companion volume to the present book, covering macroscopic thermodynamics. The author is still very happy with it and recommends it to one and all.

Barrow, G. M., *Molecular Spectroscopy*, McGraw-Hill Book Co., New York, 1962. This book emphasizes the physical picture of the energy levels of molecules and is eminently suited for beginners.

Davidson, N., *Statistical Mechanics*, McGraw-Hill Book Co., New York, 1962. A large, modern book applying statistical mechanics to a variety of technical problems.

Eisberg, R. M., *Fundamentals of Modern Physics*, John Wiley & Sons, New York, 1961. This admirable text gives an excellent presentation of basic quantum mechanics. Most of the description and some of the mathematics are suitable for beginners.

Eyring, H., D. **Henderson**, B. J. **Stover, and** E. M. **Eyring**, *Statistical Mechanics and Dynamics*, John Wiley & Sons, New York, 1964. A large, modern book applying statistical mechanics to a variety of advanced problems.

Fowler, R., **and** E. A. **Guggenheim**, *Statistical Thermodynamics*, Cambridge University Press, Cambridge, 1939. A large book applying statistical mechanics to a variety of problems. Formidable notation, but some discussions more thorough than found elsewhere.

Gibbs, J. W., *Elementary Principles in Statistical Mechanics*, Yale University Press, New Haven, Conn., 1902. The great work in which Gibbs ordered and unified statistical mechanics. A source of inspiration and ideas, not a text from which to learn the fundamentals.

Heitler, W., *Elementary Wave Mechanics*. Oxford University Press, London, 1945. This little book concentrates on giving a physical picture of quantum mechanics. Suitable for beginners.

Hill, T. L., *Introduction to Statistical Thermodynamics*, Addison-Wesley Publishing Co., Reading, Mass., 1960. A large, modern book applying statistical mechanics to a variety of technical problems.

Hirschfelder, J. O., C. F. **Curtiss, and** R. B. **Bird**, *Molecular Theory of Gases and Liquids*, John Wiley & Sons, New York, 1954. A valuable reference for studying the application of statistical mechanics to all kinds of substances. Contains a great deal on the nonequilibrium statistical mechanics of gases.

Janz, G. J., *Thermodynamic Properties of Organic Compounds*, Academic Press, New York, rev. ed., 1967. Gives a large number of ways of estimating properties of complicated compounds from their known structure through statistical mechanics. An important book in aiding one's physical "feel" for the behavior of organic compounds.

Karplus, M., **and** R. N. **Porter**, *Atoms & Molecules*, W. A. Benjamin, New York, 1970. A thorough, well-written text on quantum mechanics, atomic and molecular structure, and molecular spectra.

Mayer, J. E., **and** M. G. **Mayer**, *Statistical Mechanics*, John Wiley & Sons, New York, 1940; 2nd ed. in preparation. A hard book containing much meat. An enormous number of scientists were brought up on this.

McQuarrie, D. A., *Statistical Thermodynamics*, Harper & Row, New York, 1973. A large, modern book applying statistical mechanics to a variety of technical problems.

Tolman, R. C., *The Principles of Statistical Mechanics*, Oxford University Press, London, 1938. A long, painstaking exposition of the fundamentals of classical mechanics, quantum mechanics, and statistical mechanics. For this reason, it is a very valuable reference, although difficult for the beginner.

APPENDIX B: DEFINITE INTEGRALS*

$$\int_{-\infty}^{\infty} dz\, e^{-az^2} = 2\int_{0}^{\infty} \cdots = 2\int_{-\infty}^{0} \cdots = \frac{\pi^{1/2}}{a^{1/2}} \tag{B-1}$$

$$\int_{0}^{\infty} dz\, z e^{-az^2} = \int_{-\infty}^{0} dz |z| e^{-az^2} = \frac{1}{2a} \tag{B-2}$$

$$\int_{-\infty}^{\infty} dz\, z^2 e^{-az^2} = 2\int_{0}^{\infty} \cdots = 2\int_{-\infty}^{0} \cdots = \frac{\pi^{1/2}}{2a^{3/2}} \tag{B-3}$$

$$\int_{0}^{\infty} dz\, z^3 e^{-az^2} = \frac{1}{2a^2} \tag{B-4}$$

*Proofs of all of these are worked out in the problems, except for Eq. B-9, for which see F. H. Crawford, *Heat, Thermodynamics, and Statistical Physics* (Harcourt Brace Jovanovich, New York, 1963), p. 680.

$$\int_{-\infty}^{\infty} dz\, z^4 e^{-az^2} = 2\int_{0}^{\infty} \cdots = 2\int_{-\infty}^{0} \cdots = \frac{3\pi^{1/2}}{4a^{5/2}} \tag{B-5}$$

$$\int_{0}^{\infty} dz\, z^5 e^{-az^2} = \frac{1}{a^3} \tag{B-6}$$

$$\int_{0}^{\infty} dz\, e^{-az} = \frac{1}{a} \tag{B-7}$$

$$\int_{0}^{\infty} dz\, z e^{-az} = \frac{1}{a^2} \tag{B-8}$$

$$\int_{0}^{\infty} \frac{dz\, z^3}{e^z - 1} = \frac{\pi^4}{15} \tag{B-9}$$

$$\frac{d}{dx}\int_{0}^{x} f(z)\, dz = f(x) \tag{B-10}$$

$$\int_{0}^{x} \ln(1 - e^{-z}) z^2\, dz = \frac{x^3}{3}\ln(1 - e^{-x}) - \int_{0}^{x} \frac{z^3\, dz}{3(e^z - 1)} \tag{B-11}$$

APPENDIX C: GLOSSARY OF SYMBOLS VALUES OF CONSTANTS

Numbers in parentheses refer to equations in which symbols are first used or thoroughly defined. Symbols that appear infrequently or in one chapter only are not listed.

A = Helmholtz free energy (8-1)
Å = angstrom unit, Å $= 10^{-8}$ cm
$B(T)$ = second virial coefficient (20-22)

C = degrees Celsius = degrees centigrade
c = velocity of light in vacuum, $c = 3.00 \times 10^{10}$ cm/sec
C_P = heat capacity of system at constant pressure (8-14)
C_V = heat capacity of system at constant volume (8-14)
D_e = depth of intermolecular potential curve (Fig. 11-2)
D_0 = energy needed to break a chemical bond (Fig. 11-2)
E = internal energy of a macroscopic system (Chap. 4)
e = base of natural logarithms; $e = 2.72$
E_i = energy of macroscopic system in its ith quantum state (4-1)
f = distribution function (2-44)
f_1 = one-particle distribution function (17-5)
f_N = complete distribution function (17-1)
G = Gibbs free energy of system (8-13)
g = acceleration of gravity, $g = 980$ cm/sec^2 at sea level
g_i = degeneracy of the ith molecular energy level
g_0 = degeneracy of the lowest molecular electronic energy level (11-42)
H = enthalpy of system (8-12)
h = Planck's constant, $h = 6.625 \times 10^{-27}$ erg-sec
$\hbar$ = Planck's constant divided by 2π, $\hbar = 1.054 \times 10^{-27}$ erg-sec
I = moment of inertia of molecule (11-2)
j = rotational quantum number (11-2)
k = Boltzmann's constant, $k = 1.38 \times 10^{-16}$ erg/K $= 1.36 \times 10^{-22}$ cm^3-atm/K
K = degrees Kelvin
K = equilibrium constant (12-8)
L = mean free path (18-26)
M = molecular weight
m = mass of a particle
N = number of particles in system
n = vibrational quantum number (11-23)
$\mathbf{n}$ = vector in quantum number space (Fig. 10-1)
N_i = number of particles of type i in system (8-7)
N_i = occupation number of particle state i (9-3)
N_0 = Avogadro's number, $N_0 = 6.02 \times 10^{23}$/mole
P = thermodynamic pressure (8-9)
p = magnitude of the vector $\mathbf{p}$ (17-1)
$\mathbf{p}$ = momentum vector for a particle (17-1)
P_i = probability that the N-particle system is in the ith quantum state (4-1)
p_x = x-component of momentum $\mathbf{p}$ (17-1)
Q = partition function for the system (5-12 and 8-6)
q = heat absorbed by system (6-1)
q = partition function for a single particle (9-29 and 11-1)

Q_N = same as Q with number of particles noted explicitly (9-3)
Q_{int} = internal contribution to Q (10-19)
R = universal gas constant, $R = 1.987$ cal/K-mole $= 8.314 \times 10^7$ erg/K-mole
$= 0.082$ liter-atm/K-mole
r = magnitude of the vector $\mathbf{r}$ (2-72)
$\mathbf{r}$ = position vector locating the center of mass of a particle (2-72)
$\mathbf{r}_{ij}$ = position of particle i relative to particle j, $\mathbf{r}_{ij} = \mathbf{r}_i - \mathbf{r}_j$
r_e = equilibrium interatomic separation (Fig. 11-2)
S = entropy of system (6-2)
s = collision cross section (Fig. 18-2)
T = absolute temperature
U_{ij} = intermolecular potential energy between particles i and j (20-1)
V = volume of system
v = magnitude of the vector $\mathbf{v}$ (17-22)
$\bar{v}$ = average speed of a particle (17-37)
$\mathbf{v}$ = velocity vector for a particle (17-1)
$\mathbf{v}_{21}$ = velocity of particle 2 relative to velocity of particle 1 (18-11)
v_{mp} = most probable speed of a particle (17-32)
v_{rms} = root mean square speed of a particle (17-43)
v_x = x-component of velocity $\mathbf{v}$
W = degeneracy of energy level for N-particle system (4-2)
w = work done on system (6-1)
x = rectangular coordinate (2-72)
y = rectangular coordinate (2-72)
z = rectangular coordinate (2-72)
Z_1 = collision frequency for a particle (18-23)
$\beta = 1/kT$ (5-9, 6-16, and 10-25)
Γ = particle flux due to effusion (18-31)
δ = deviation from the mean (operator) (2-30)
ϵ = energy of a small part of the system (5-2)
ϵ_{trans} = translational energy of a particle (10-5)
θ = one of the spherical coordinates (2-72)
Θ = characteristic temperature of degeneration (14-1)
Θ_D = Debye characteristic temperature (13-29)
Θ_E = Einstein characteristic temperature (13-7)
Θ_{rot} = characteristic temperature of rotation (11-3)
Θ_{trans} = characteristic temperature of translation (10-12)
Θ_{vib} = characteristic temperature of vibration (11-24)
μ = molecular chemical potential or molecular Gibbs free energy (8-20)
ν = frequency (cycles/sec), $\nu = \omega/2\pi$
$\tilde{\nu}$ = wave number, $1/\lambda$, of radiation
ν_i = stoichiometric number of species i (12-2)

ρ = particle density, particles/unit volume, $\rho = N/V$ (17-14)
σ = symmetry number for rotation (11-6)
Φ = distribution function for one component of velocity (17-21)
ϕ = one-particle distribution function for molecular speeds (17-23)
ϕ = one of the spherical coordinates (2-72)
Φ_1 = one-particle velocity distribution function (17-20)
φ_1 = one-particle momentum distribution function (17-15)
ω = angular velocity in (rad/sec), sometimes called *frequency*, $\omega = 2\pi\nu$

APPENDIX D: UNIT CONVERSIONS AND VALUES OF CONSTANTS

GLOSSARY

A	ampere
C	coulomb
Ω	ohm
J	joule
W	watt
V	volt
eV	electron volt

Energy	erg = dyne-cm = g-cm^2/sec^2
	J = 10^7 erg = W-sec = V-C
	cal = 4.1840×10^7 erg = 4.184 J = 41.29 cm^3-atm = 0.04129 liter-atm.
	cm^{-1} = 1.986×10^{-16} erg/molec = 2.859 cal/mole
	eV = 1.602×10^{-12} erg = 23,060 cal/mole
Gas constant	R = 1.98726 cal/K-mole = 0.082059 liter-atm/K-mole = 8.3147 J/deg-mole

Boltzmann's constant	$k \equiv R/N_0 = 1.3804 \times 10^{-16}$ erg/K $= 3.2994 \times 10^{-24}$ cal/K
Avogadro's number	$N_0 = 6.02308 \times 10^{23}$ mole^{-1}
Mass	kg = 2.2046 lb
Gravity	$g^0 = 980.665$ cm/sec^2
Electricity	C = A-sec
	J = V-C
	V = Ω-A
	$\mathcal{F} = 96{,}493$ C/eq = 23,062.4 cal/V-eq
	$e \equiv \mathcal{F}/N_0 = 1.6020 \times 10^{-19}$ C
	kW-hr = 3.600×10^6 J
Mathematics	$e = 2.7183$
	$\ln x = 2.303 \log x$
	$\pi = 3.14159$
Pressure	atm = 1,013,250 dyne/cm^2
	atm = 760 mm Hg = 760 torr
Light	$c = 2.9979 \times 10^{10}$ cm/sec
Planck's constant	$h = 6.625 \times 10^{-27}$ erg-sec
	$\hbar = 1.0544 \times 10^{-27}$ erg-sec
Length	in = 2.5400 cm

APPENDIX E: ANSWERS AND HINTS TO PROBLEMS

1-2. The author obtained 2×10^{17}.

2-1. $f(x) = 1/L$ $(x \leqslant L)$, $f(x) = 0$ $(x > L)$; $\frac{1}{2}L$; $1/\sqrt{3}$.

2-2. $f(t) = (2/25)$ week$^{-1} - 2/(25)^2$ week^{-2}t for $(0 \leqslant t \leqslant 25$ week); 25/3 week; 7.32 week; 114 rats.

2-3. (*a*) 0.75, (*b*) 0.0138.

2-4. $f(v) = 4\pi\,(m/2\pi kT)^{3/2} v^2 \exp(-mv^2/2kT)$; $\bar{v} = (8kT/\pi m)^{1/2}$.

2-5. $$f(v_x,v_y,v_z)=(m/2\pi kT)^{3/2}\exp\left[-m(v_x^2+v_y^2+v_z^2)/2kT\right].$$

No. $f(v_x)=(m/2\pi kT)^{1/2}\exp(-mv_x^2/2kT)$.

$$\bar{v}_x=0=\bar{v}_y=\bar{v}_z. \qquad \overline{v_x^2}=kT/m;\ \overline{\tfrac{1}{2}mv_x^2}=\tfrac{1}{2}kT.$$

2-7. $f(w,A)=(1600\text{ lb-pt})^{-1}$ in the range; $\overline{V/w}=0.022$ pt/lb. Not correlated in range where $f\neq 0$, but that range establishes correlation.

2-8. $\frac{7}{6}v_0$; all between v_0 and $2v_0$ equally probable; $\frac{1}{3}$.

2-9. $P(i \text{ or } j)=[n_i+n_j-n(i \text{ and} j)]/n=P_i+P_j-P(i \text{ and } j)$.

2-11. $P(i|j)=P_i$; imposing the condition does not change the probability.

2-12. $P_{ij}=P_iP(j|i)=P_jP(i|j)$.

2-13. $$f_2(x,z)=\int dw\,dy\,f_4(w,x,y,z);\quad f_3(w,x,y)=\int dz\,f_4(w,x,y,z);$$

$$\bar{h}=\int dw\,dx\,dy\,dz\,h(w,z)f_4(w,x,y,z)=\int dw\,dz\,h(w,z)f_2(w,z).$$

2-14. From P_{ij} one may obtain not only P_i and P_j but also $P(i|j)$ and $P(j|i)$. Thus P_{ij} gives the relative number of members of the ensemble having both i and j simultaneously, whereas P_i and P_j together do not. They can *never* furnish as much information as P_{ij} because even when i and j are uncorrelated, that fact is not implied by the P_i and P_j unless $P_{ij}=P_iP_j$ is given.

2-16. $\overline{\ln P}=\Sigma_i P_i \ln P_i$; $\overline{\ln P}=0$ when all members in same state. That is its largest value. As ensemble becomes less clustered, $\overline{\ln P}$ decreases (i.e., becomes increasingly negative).

3-2. $\frac{3}{2}NkT=10^{-3}$ cal; $N=7\times10^{17}$.

4-3. $P(2h)=\frac{1}{4}$, $P(2t)=\frac{1}{4}$, $P(1h \,\&\, 1t)=\frac{1}{2}$; $P(4h)=\frac{1}{16}$, $P(4t)=\frac{1}{16}$, $P(2h \,\&\, 2t)=\frac{3}{8}$; $P(6h)=\frac{1}{64}$, $P(6t)=\frac{1}{64}$, $P(3h \,\&\, 3t)=\frac{5}{16}$.

6-4. See L. Pauling, *The Nature of the Chemical Bond* 2nd ed. (Oxford University Press, London, 1940), pp. 301–303.

7-1. 2.585 bit, 5.170 bit, 4.337 bit. In the first case, the initial field of uncertainty is six numbers of probability 1/6. In the second, this field is reduced twice. The initial field in the third has six pairs with probability 1/36 and 15 different unlike pairs with probability 1/18.

7-2. 0.0633 bit.

7-3. (1) 0.193 bit, (2) 1.000 bit, (3) 1.415 bit, (4) 3.000 bit; (1) 0.193 bit,

(2) 0.807 bit, (3) 1.000 bit, (4) 1.000 bit, total is 3.000 bit, which is the same, because the total information does not depend on order in which information is conveyed.

7-4. $I = \text{bit} \ln(L/\Delta x)/\ln 2$. $\Delta N = k \ln(L/\Delta x)$. Infinite information and infinite negentropy. Since the amount of energy that must be degraded into heat must be at least $T\Delta N$, something like the uncertainty principle seems reasonable.

7-6. Since 2 bits per letter, there must be about $\frac{1}{2} \times 10^{10}$ nucleic acid residues in such a sperm cell.

7-7. $I = 4.32$ bits/letter. $n = 4.32/2 = 2.16$ nucleotide pairs. Easily consistent.

7-8. Does not treat importance of information, cannot cope with novelty, cannot cope with purpose.

8-1. $Q = 1 + 2e^{-\beta\epsilon}$; $A = -kT \ln Q$; etc. As T goes to zero, the system goes into its ground state (i.e., $P_1 = 1, P_2 = P_3 = 0$). As T becomes large with respect to ϵ/k, the probabilities of the states become equal (i.e., the P's tend to $\frac{1}{3}$). At low T, E is 0, at high T, E is $2\epsilon/3$, the increase occurs at about $T = \epsilon/k$. The maximum rate of increase occurs at the maximum in the heat capacity, which is near 0 for low T and for high T, but shows a peak at what value of T? Such humps in heat capacity curves are characteristic of the sudden thawing out of energy levels. They have often been observed experimentally and are called the *Shottky anomaly*.

9-1. Fermions: states: T–H only; therefore $P(H\text{–}T) = 1$, $P(H\text{–}H) = P(T\text{–}T) = 0$. Bosons: states: T–T, H–T, T–T; thus $P(H\text{–}T) = P(H\text{–}H) = P(T\text{–}T) = \frac{1}{3}$. Distinguishable: states: H–H, H–T, T–H, T–T; thus $P(H\text{–}T \text{ or } T\text{–}H) = \frac{1}{2}$, $P(H\text{–}H) = P(T\text{–}T) = \frac{1}{4}$. Boltzmann statistics gives the same probabilities as for distinguishable. It differs in using a different partition function for calculating thermodynamic variables: It would be inappropriate in this case.

9-2. $\mu < 0$. Hint: N_i must be positive for all states i, even in the case of the ground state.

9-3. 1.17.

9-4. How can ϵ_i be proportional to $V^{2/3}$ for more than one choice of energy zero?

10-1. $\text{Ratio} = e^{(3\pi^2\hbar^2/2ma^2kT)} = e^{(9\times 10^{-19})} \approx 1$. The energy difference between successive levels is infinitesimal compared to kT. Note that three states have the higher level and only one has the lower, thus there are three times as many particles in the higher level as in the lowest.

10-2. Experimental studies of He, Ar, Hg, and Cd vapors confirm this result. The first derivations were by O. Sackur, *Ann. Physik*, **36**, 958 (1911); **40**, 67 (1913). H. Tetrode, *Ann. Physik*, **38**, 434 (1912).

10-3. See Table 11-2.

10-4. About 4 K, which is approximately the liquefaction temperature for He. The T would be about 0.66 times as large.

10-8. Hint: let the desired integral $I = \int_{-\infty}^{\infty} dx\, e^{-ax^2}$. Since the name given the dummy integration variable is unimportant, $I = \int_{-\infty}^{\infty} dy\, e^{-ay^2}$ also. Thus $I^2 = \int_{-\infty}^{\infty}\int_{-\infty}^{\infty} dx\, dy\, e^{-a(x^2+y^2)}$. This can be viewed as an integral over a two-dimensional surface and integrated easily after being put into polar coordinates. Knowledge of this trick is a source of satisfaction to many people.

11-1. (*a*) 1, (*b*) 3, (*c*) 7, (*d*) 9.

11-2. Ratio $= e^{\Theta_{\text{vib}}/T}$. In this case the ratio $= e^{3340/298} = 7.4 \times 10^4$. Note: this is the constant ratio of populations for *each* successive vibrational level to the one below it.

11-3. Fraction $= \int_{z_0}^{\infty} dz\, e^{-z\Theta_{\text{rot}}/T}/\sigma q_{\text{rot}}$, where $z_0 = 2IkT/\hbar^2$. Thus the fraction is e^{-1}.

11-4. Ratio $= e^0/[(2+1)e^{-(1)(2)\Theta_{\text{rot}}/T}] = e^{2\Theta_{\text{rot}}/T}/3$. In this case, the ratio is .034. Note: this is *not* a constant ratio.

11-8. $k\Theta(6e^{-2\Theta/T} + 30e^{-6\Theta/T} + 84e^{-12\Theta/T})/(1 + 3e^{-2\Theta/T} + 5e^{-6\Theta/T} + 7e^{-12\Theta/T})$.

11-10. Hint: in expanding the exponential in the denominator of the second term in Eq. 11-31, three terms must be taken.

11-11. $\ln Q = N \ln(2 + 2e^{-178/T})$, from which the properties follow. The contribution to P is zero.

11-12. Starts at $\frac{3}{2}Nk$, but at about 2 K rises to $\frac{5}{2}Nk$, where it has a plateau, finally rising to $\frac{7}{2}Nk$ at about 3000 K.

11-13. For CN, $q_{\text{rot}} = 204.5$. This follows from $q_{\text{rot}} = 2M_1M_2r^2kT/\sigma\hbar^2N_0(M_1 + M_2)$.

11-18. Slide rule accuracy should yield about 30 cal/K-mole. If the most precise values of all constants are used, the result is about 30.126 cal/K-mole, which is precisely the experimental value.

11-20. A review of the argument leading to $Q = q^N/N!$ for one-component gases will show the way to a proof that in this case $Q = q_a^{N_a} q_b^{N_b}/N_a!N_b!$.

11-21. Different gases: $\Delta S = N_a k \ln[(V_a + V_b)/V_a] + N_b k \ln[(V_a + V_b)/V_b]$. Same gases: $\Delta S = N_a k \ln[(V_a + V_b)N_a/(N_a + N_b)V_a] + N_b k \ln[(V_a + V_b)N_b/(N_a + N_b)V_b]$. Note: The "entropy of mixing," as usually defined in thermodynamics, is based on the indistinguishability of identical molecules, whose interchange does not lead to a new quantum state.

11-22. In complete analogy with the three-dimensional case: $Q_{\text{trans}} = q_{\text{trans}}^N/N! = (mkT\mathcal{A}/2\pi\hbar^2)^N/N!$. $\mathcal{P} = kT(\partial \ln Q/\partial \mathcal{A})_{T,N}$; $\mathcal{P}\mathcal{A} = NkT$.

11-23. $q_{\text{elect}} = 2\sum_{n=1}^{\infty} e^{\beta a/n^2} < 2\sum_{n=1}^{\infty} 1 = \infty$. Thus $q_{\text{elect}} = \infty$, and $e^{\beta a/n^2}/q_{\text{elect}} = 0$ for *any* state, *even the ground state*. This is delightful. Hint: see Hirschfelder, Curtiss, and Bird, pp. 268–271. Also see Prob. 12-4.

11-24. Trans: $\frac{3}{2}Nk$, rot: $\frac{3}{2}Nk$, internal rotation: $\frac{2}{2}Nk$, total $C_V = 4Nk$.

11-25. $1:3:5$; $1:10^{-6}:10^{-12}$. 1800 K.

11-30. 0.210, 6.06×10^{-10}.

12-1.

$$\begin{aligned}
\rho_A\rho_B/\rho_{AB} &= (q_A/V)(q_B/V)/(q_{AB}/V)\\
&= \left\{(m_AkT/2\pi\hbar^2)^{3/2} g_{0,A}(m_BkT/2\pi\hbar^2)^{3/2} g_{0,B}\right\}\\
&\div \left\{[(m_A+m_B)kT/2\pi\hbar^2]^{3/2}(T/\Theta_{\text{rot}})[e^{-\Theta_{\text{vib}}/2T}/\right.\\
&\quad \left.(1-e^{-\Theta_{\text{vib}}/T})]\, g_{0,AB}e^{(D_0/kT+\Theta_{\text{vib}}/2T)}\right\}\\
&= [m_Am_BkT/(m_A+m_B)2\pi\hbar^2]^{3/2}(\Theta_{\text{rot}}/T)(1-e^{-\Theta_{\text{vib}}/T})\\
&\quad \times (g_{0,A}g_{0,B}/g_{0,AB})e^{-D_0/kT}.
\end{aligned}$$

Note particularly how the dissociation energy enters the denominator in q_{zero} as a correction for the different zero of energy in the conventional q_{AB}. For temperatures below Θ_{vib}, this is proportional to $\sqrt{T}\,e^{-\epsilon/kT}$, thus increases with temperature. For very high temperatures this becomes proportional to $T^{-1/2}$, thus decreases with temperature, which is unreasonable. The assumption of an infinite number of evenly spaced vibrational levels of course breaks down. There can be only a finite number of levels in a potential well of finite depth. The same holds for rotational levels, because if too much rotational energy is present, centrifugal force will break the chemical bond.

12-2. Essentially identical, except that the symmetry number, 2, for A_2 makes the equilibrium constant twice as large.

12-3. $\rho_{A^+}\rho_{e^-}/\rho_A \approx (2g_{0,A^+}/g_{0,A})(m_{e^-}kT/2\pi\hbar^2)^{3/2}e^{-\epsilon^\circ/kT}$.

12-4. Let $x = T/\Theta_{\text{ion}}$. Then $10^{20}\ \text{cm}^{-3} \approx 2(m_e - \epsilon^\circ/2\pi\hbar^2)^{3/2}x^{3/2}e^{-1/x}$; $e^{1/x} \approx 2470\, x^{3/2}$, the solution of which is about 0.16. This is most easily found by letting $z = 1/x$ and solving $z = 9.02 - 3.45\log z$ by trial and error. Thus most of the atoms are ionized at temperatures small compared with Θ_{ion}. Since excited *electronic* levels in most atoms are of the same order as $\epsilon°$, the use of ground electronic energies is justified even under conditions of extreme ionization! This is especially true at densities lower than the rather high $10^{20}/\mathrm{cm}^3$ chosen for this problem. See Prob. 11-23.

12-5. $\alpha=\rho_{H^+}/\rho_{H^0}\approx\rho_{H^+}/\rho_H=\rho_{e^-}/\rho_H;\ \rho_{H^+}\rho_{e^-}/\rho_H\approx\rho_{H^0}\alpha^2=(2g_{0,A^+}/g_{0,A})(m_{e^-}kT/2\pi\hbar^2)^{3/2}e^{-\epsilon^0/kT};\ \alpha=7.7\times10^{-3}$.

12-6. $K=(m_{HCl}/m_{H_2})^{3/2}(m_{HCl}/m_{Cl_2})^{3/2}(4\Theta_{rot,H_2}\Theta_{rot,Cl_2}/\Theta^2_{rot,HCl})\times\{\exp[(2D_{0,HCl}-D_{0,H_2}-D_{0,Cl_2})/kT]\}\{[1-\exp(-\Theta_{vib,H_2}/T)]\times[1-\exp(-\Theta_{vib,Cl_2}/T)]/[1-\exp(-\Theta_{vib,HCl}/T)]^2\}\approx6.3\times10^9$.

12-12. (37–37): $\frac{1}{16}$; (35–35): $\frac{9}{16}$; (35–37): $\frac{6}{16}$. $S=(1.987\text{ eu}/16)\ln 16+[(9)\times(1.987\text{ eu})/16]\ln(16/9)+[6(1.987\text{ eu})/16]\ln(16/6)=1.718$ eu. Not at all, since the different choice of standard condition will lead to a different value of the entropy. In one case these are the pure separated isotopes, in the other case the naturally occurring mixture.

12-13. $K_P=0.259\text{ dyne/cm}^2=2.56\times10^{-7}$ atm.

12-14. $K_P=3.25$.

12-15. $K_P=4.0$.

12-16. The literature value is 6.9×10^7, with no units given. What sense does that make?

12-17. 1/3 are dissociated. Since 0.2 $D_0=16.28\times10^{-13}$ erg, and $h\nu=3.08\times10^{-13}$, at least 6 quanta of vibrational excitation are needed; thus the fraction is of the order of $e^{-6\Theta_{vib}/T}=0.028$. The entropy effect favors dissociation. In terms of dynamics this means that although *very* rarely does a molecule obtain enough energy to dissociate, the density is so low that once dissociated, the atoms *almost never* collide with one another to recombine to form molecules.

12-19. One can solve simultaneously the following: $X(A_2)+X(B_2)+X(AB)=1$; $[2X(A_2)+X(AB)]/[2X(A_2)+2X(B_2)+2X(AB)]=z$; $X(AB)^2/X(A_2)X(B_2)=4$. The results are $X(A_2)=z^2$; $X(B_2)=(1-z)^2$; $X(AB)=2z(1-z)$; independent of T and P.

12-20. Since ln K is off by ±4.58, K is off by a factor of about 100.

13-3. $\alpha=(\partial\Theta_D/\partial P)_T[12NkD(\Theta_D/T)/V\Theta_D-9Nk/(VTe^{\Theta_D/T}-VT)]$. High temp: $\alpha=(\partial\ln\omega_{max}/\partial P)_T C_V/V$. Hints: recall one of the Maxwell relations from thermodynamics; use Eqs. 13-37 and B-10.

13-12. Equate the chemical potentials for solid and gas, correcting the former by $-\Delta h_s$ for the difference in energy zeros. Use the ideal gas law for gas-phase pressure, which proves to be $P=(m/2\pi\hbar^2)^{3/2}(kT)^{5/2}(1-e^{-\Theta_E/T})^3 e^{3\Theta_E/2T} e^{-\Delta h_s/kT}$.

14-1. Zero. Perhaps the best way is to use Eq. 6-26, with $W=1$ for the N-particle state having the lowest-lying N particle-states occupied.

14-2. Since μ is simply a measure of the energy change on adding one more particle to the system, and at 0 K that particle must have energy ϵ_{max}.

14-3. Equate the electron heat capacity to $3Nk$, since obviously are well within the Dulong-Petit region. Result: $T=56{,}000$ K.

14-4. Plug into Eq. 14-2, get $\Theta=36{,}900$ K. The volume makes no difference at all, since V appears in Eq. 14-1 only in the form N/V and the density remains constant.

14-5. Electron density is same as atomic density $=5.68\times10^{22}$ cm^{-3}. Use Eq. 14-8, note that m is the mass of a single electron, $\mu_0=8.70\times10^{-12}$ erg. From this, $\Theta=\mu_0/k$, or 63,100; thus the melting point is at 0.02 of Θ, and the degenerate electron gas is an excellent model at all temperatures up to the melting point.

15-3. Compare Eq. 15-4 with Eq. 15-3. Photons greater: $\omega<0.69kT/\hbar$; states greater: $\omega>0.69kT/\hbar$. For $\omega\gg0.69kT/\hbar$, the unity in the parentheses of the denominator of Eq. 15-5 can be neglected. The result then is the same as the use of Boltzmann statistics would give. Note: for radiation, *low* temperatures give an appreciable fraction of frequencies satisfactorily described by Boltzmann statistics.

15-4. $P(25\text{ C})=1.9\times10^{-11}$ atm. $T(P=1\text{ atm})=1.43\times10^5$ K.

15-5. $T=8660$ K.

16-1. $V=4.09\times10^{-20}\text{cm}^3$. $\overline{\delta N^2}=N=1$; thus $(\overline{\delta N^2})^{1/2}=1$ also.

16-2. $\overline{\left[(\delta F)^2\right]^{1/2}}=\overline{|\delta F|}$ and is smaller because greater values receive less weight in the averaging.

16-3. $\overline{\delta E^2}/\bar{E}^2=20\Theta_D^3/3\pi^4NT^3$. For carbon, $T=2.47\times10^{-4}$ K for 1 mg. Since diamond has the highest value of Θ_D, other substances will have still lower temperatures. Hint: see Eq. 5-23.

16-4. Hint: review Eqs. 5-23 and 5-24. Recall that $\bar{E}=-(\partial Q/\partial\beta)/Q$; $\overline{E^2}=(\partial^2Q/\partial\beta^2)/Q$; $\overline{E^3}=-(\partial^3Q/\partial\beta^3)/Q$.

16-5. This is a good example of a case in which knowledge of T does not imply knowledge of E. Hint: Gibbs (p. 75) was the first to realize that this result both followed from statistical mechanics and was true experimentally.

16-6. Hints: how does μ change when particles are added to a two-phase system? How are the fluctuations observed? Note: also, for a system at its *critical point*, addition of more particles does not change μ, and widespread fluctuations are expected. The extraordinary light-scattering ability of liquids at their critical points attests to this.

17-4. Prove Eq. B-2 by substituting $x = z^2$. Prove Eq. B-4 by considering the derivative with respect to a of Eq. B-2.

17-5. Consider derivatives with respect to a of Eq. B-1.

17-6. $v_{1,\mathrm{rms}}/v_{2,\mathrm{rms}} = \bar{v}_1/\bar{v}_2 = v_{1,\mathrm{mp}}/v_{2,\mathrm{mp}} = (m_2/m_1)^{1/2}$. In this case, answer is 4.

17-7. 1.78×10^5 cm/sec, 4.75×10^4 cm/sec. These are of the same orders of magnitude as the velocities of sound in these gases.

17-8. Hint: Consider what is meant by $\phi(v)\,dv$. Answer: $2\pi(\pi kT)^{-3/2}e^{-\kappa/kT}\kappa^{1/2}\,d\kappa$. $\kappa_{\mathrm{mp}} = \frac{1}{2}kT$. Why is $\frac{1}{2}mv_{\mathrm{mp}}^2$ not equal to κ_{mp}?

17-9. What if the system were translating with respect to the origin of the velocity coordinate system?

17-10. Hint: The equation $\lambda = \lambda_0(1 + v_z/c)$ can be solved for v_z. Since the observer sees from only one direction, one starts with Eq. 17-21, using the fact that $\Phi(v_z)\,dv_z$ times I is the radiation seen by molecules whose z-component of velocity lies between v_z and $v_z + dv_z$. From this, the proof is easy that $I(\lambda) = (Ic/\lambda_0)(m/2\pi kT)^{1/2}e^{-mc^2(\lambda-\lambda_0)^2/2\lambda_0^2kT}$. This is sometimes written in the form $I(\lambda) = I(\lambda_0)e^{-mc^2(\lambda-\lambda_0)^2/2\lambda_0^2kT}$. Why? Very high temperatures are often measured by the Doppler widths of spectral lines emitted from very hot gases.

17-11. $2y/\sqrt{\pi}$, 1, erfc $x = 1 -$ erf x, erf $\frac{1}{2}$.

17-12. Not at all. The high-energy tails would be lowered and would tend to zero at a speed such that a single particle had all the energy available to the system.

17-13. 0.349 km/sec compared with 0.421 km/sec.

18-1. 1.0×10^{10} sec^{-1}.

18-2. 1.25×10^{29} sec^{-1}cm^{-3}.

18-3. $L = 4.54\times10^{-6}$ cm; $L/r = 207$.

18-6. $N = 3.22\times10^5$, $L = 7300$ m.

18-7. H_2 effuses twice as fast into the O_2 as O_2 effuses into the H_2, despite the higher oxygen pressure. The total pressure in the O_2 side increases. Of course this is only temporary, since at equilibrium each species will have its partial pressure the same on the two sides.

18-8. 1.8×10^{18}.

18-10. $P = (\Delta m/\pi r^2 t)(2\pi RT/M)^{1/2} = 2.10\times10^{-4}$ atm. Ratio $= 12$; it should be $\ll 1$, but these equations are used anyway.

18-11. The total number of collisions per unit area per unit time is given by Eq. 18-29. In spherical coordinates that equation is

$$\Gamma = \frac{N}{mV}(2\pi mkT)^{-3/2}\int_0^{\infty} p^2\,dp\int_0^{\pi/2}\sin\theta\,d\theta\int_0^{2\pi} d\chi\, p\cos\theta\, e^{-p^2/2mkT}.$$

With no integrals at all, this is number of collisions per unit area per unit time in which the colliding particles have p between p and $p+dp$, θ between θ and $\theta+d\theta$, and χ between χ and $\chi+d\chi$. The quantity requested is precisely this, integrated over p and χ only.

18-12. $(\pi N/V)(m/2\pi kT)^{3/2}v^3 e^{-mv^2/2kT}\,dv$. See the hint for Prob. 18-11.

18-13. The number of effusing molecules is given by Eq. 18-29. The total v^2 for effusing molecules is given by a similar expression with v^2 added to the integrand. The average of v^2 is the quotient of these two integrals. $v_{\text{rms}}(\text{effuse}) = 2(kT/m)^{1/2}$; $KE = 2kT$, compared with $\frac{3}{2}kT$. Since faster molecules are more likely to effuse through the hole than slow ones, the average speed of the molecules that pass through is greater than the average speed of all particles in the system.

18-14. Pick a particle in each member of the ensemble and follow it. Likelihood of its colliding in dl is proportional to dl (i.e., to the volume $s\,dl$ swept out, times ρ). Let N be the number of members of ensemble (whose total number $= N_0$) that have not yet collided: $dN = -\alpha N\,dl$, $dN/N = -\alpha\,dl$; $N/N_0 = P(\text{not yet collided}) = e^{-\alpha l}$. Now to prove that α is $1/L$, we know that $\alpha N\,dl = \alpha N_0 e^{-\alpha l}\,dl$ is the number of collisions within dl; thus the probability of a collision in dl is $\alpha e^{-\alpha l}\,dl$, and the average of $l = L = \int_0^{\infty} \alpha l\, e^{-\alpha l}\,dl$ $= 1/\alpha$. Note: nothing was said about the past history of the particles chosen to follow. Thus not only have we found the distribution of distances between collisions, we have also found the distribution of free paths for *any* group of gas molecules at equilibrium, just before or just after they are selected for whatever reason.

18-15. Problem 18-14 shows that the mean free path of particles striking wall is just L. Average distance from wall is thus $L\,\overline{\cos\theta}$, and $\overline{\cos\theta}$ is found from the result of Prob. 18-11, using Eq. 18-31 for normalization.

18-17. One first proves Eq. 18-15 valid for differing masses, where $M = m_A + m_B$, $\mu = m_A m_B/(m_A + m_B)$, $\mathbf{v}_{BA} = \mathbf{v}_B - \mathbf{v}_A$, $\mathbf{V}_{BA} = (m_A\mathbf{v}_A + m_B\mathbf{v}_B)/(m_A + m_B)$. One then proves $m_A v_A^2 + m_B v_B^2 = \mu v_{BA}^2 + MV_{BA}^2$. Then $\bar{v}_{BA}$ can be found, as in Eq. 18-22, to be $(8kT/\pi\mu)^{1/2}$. The entire procedure is parallel to Eqs. 18-12 to 18-22.

18-18. $Z_{AA}=2s_{AA}\rho_A^2(kT/\pi m_A)^{1/2}$; $Z_{BB}=2s_{BB}\rho_B^2(kT/\pi m_B)^{1/2}$; $Z_{AB}=s_{AB}\rho_A\rho_B(8kT/\pi\mu)^{1/2}$.

18-19. From the answer to Prob. 18-12 we can find $\bar{v}$ for those which effuse:

$$\bar{v}=\int_0^\infty v^4e^{-mv^2/2kT}\,dv\Big/\int_0^\infty v^3e^{-mv^2/2kT}\,dv=(9\pi kT/8m)^{1/2}$$

Thus $\bar{v}$ for those that effuse is $3\pi/8$ or 1.178 times $\bar{v}$ for all the molecules. 9.72×10^4 cm/sec.

18-20. The exact approach is to solve

$$0.1=\int_{(1-\xi)\bar{v}}^{(1+\xi)\bar{v}} v^3e^{-mv^2/2kT}\,dv\Big/\int_0^\infty v^3e^{-mv^2/2kT}\,dv.$$

The definite integral can be solved in closed form after the substitution $z=v^2$. The value of ξ that solves the resulting equation can be found by trial-and-error substitution, especially simple on an electronic slide rule, to be 0.047. An approximate approach, which gives the same answer, is to assume ξ to be so small that the integral in the numerator above can be approximated by the area of the rectangle of the same height, namely, $\bar{v}^3e^{-m\bar{v}^2/2kT}(2\xi\bar{v})$.

19-1. -0.0114 atm.

19-2. 57 km. 10^{39}.

19-3. $\frac{1}{2}=e^{-29gh/RT}$; $h=5500$ m. Note: the *average* height of a particle is harder to find and is about 8000 m. If the correct gravitational potential were used, it would be infinity! (See Prob. 19-8 for an elaboration of that.)

19-4. $L=51.3$ m.

19-5. Hint: $\rho(r)=Ae^{m\nu^2r^2/2kT}$, which needs only be normalized to give N molecules in the volume V. Answer: $\rho(r)=\dfrac{Nm\nu^2e^{m\nu^2r^2/2kT}}{2\pi kTl(e^{m\nu^2R^2/2kT}-1)}$.

19-6. 6.9×10^{23} mole^{-1}.

19-7.

$$\bar{y}=\frac{\int_0^h y\rho(y)\,dy}{\int_0^h \rho(y)\,dy}=\frac{RT/Mg-e^{-Mgh/RT}(h-RT/Mg)}{1-e^{-Mgh/RT}}$$

Note: this has a limit of 0 as $Mg/RT\gg h$, a limit of $\frac{1}{2}h$ as $Mg/RT\ll h$, and a limit of RT/Mg as $h\to\infty$. Check these three assertions for accuracy. How large is RT/Mg for air at 0 C?

19-8. $\rho(r) = \alpha e^{-m_1 m_2 G/rRT}$, where $\alpha = 1/\infty = 0$. This means that there will be a finite concentration of atmosphere at no finite height above the planet. The gravitational attraction is too weak to permit any planet to retain its atmosphere forever at a finite temperature. Thus the kinetics of the loss of a planet's atmosphere becomes most important to the development of life on that planet.

20-3. Hirschfelder, Curtiss, and Bird (p. 590) quote $\sigma = 2.4$ Å, $K = 3.7/2.4$, and $\epsilon = 14 \times 10^{-15}$ erg for neon, derived from data on viscosity.

20-4. Hint: consider $\ln \Pi$ and recall the expansion $\ln(1-x) = -x$ for small x.

20-7. Hint: Use van der Waals' equation and also the fact that the critical point is marked by an inflection point on an isotherm; thus both $(\partial P/\partial \rho)_T$ and $(\partial^2 P/\partial \rho^2)_T$ can be set equal to zero. See Andrews, p. 119.

20-8. From Eq. 20-19, $d\mu = \rho^{-1} dP$; $\Delta\mu = 0 = \int \rho^{-1} dP$. If the figure is turned on its side, it is clear that this criterion is precisely the same as the equal area rule. Since $P = P(\rho, T)$ is analytic, all paths between the same two points must give $\Delta\mu = 0$. Therefore, the result is in no way discredited because the path used in the figure utilizes unstable states (those with negative compressibility).

20-12. What relationship holds between P and V for a dense gas when its second virial coefficient is zero?

20-15. Hint: eliminate T from the equations $(\partial P/\partial d)_T = 0$ and $(\partial^2 P/\partial d^2)_T = 0$. The resulting transcendental equation for d_{crit} can be solved by trial and error, a process that is facilitated by use of an electronic slide rule.

INDEX